Springer Oceanography

The Springer Oceanography series seeks to publish a broad portfolio of scientific books, aiming at researchers, students, and everyone interested in marine sciences. The series includes peer-reviewed monographs, edited volumes, textbooks, and conference proceedings. It covers the entire area of oceanography including, but not limited to, Coastal Sciences, Biological/Chemical/Geological/Physical Oceanography, Paleoceanography, and related subjects.

More information about this series at http://www.springer.com/series/10175

Konstantin Pokazeev · Elena Sovga ·
Tatiana Chaplina

Pollution in the Black Sea

Observations about the Ocean's Pollution

Konstantin Pokazeev
Faculty of Physics
M. V. Lomonosov Moscow State University
Moscow, Russia

Elena Sovga
Shelf Hydrophysics Department
Marine Hydrophysical Institute of RAS
Sevastopol, Russia

Tatiana Chaplina
Institute for Problems in Mechanics
of the RAS
Moscow, Russia

ISSN 2365-7677 ISSN 2365-7685 (electronic)
Springer Oceanography
ISBN 978-3-030-61894-0 ISBN 978-3-030-61895-7 (eBook)
https://doi.org/10.1007/978-3-030-61895-7

This Springer imprint is published by the registered company Springer Nature Switzerland AG
The registered company address is: Gewerbestrasse 11, 6330 Cham, Switzerland

Introduction

The World Ocean is a huge but very fragile system. This has become particularly evident in recent decades, when ocean water pollution has reached unprecedented levels. Meanwhile, not only the well-being of ocean ecosystems depends on the state of ocean waters, but human civilization itself is largely dependent on the ocean since it influences weather and determines the climate of the entire planet. That is why ocean pollution has become one of the most serious environmental problems of our time.

As a result of human activities, a huge amount of pollutants enters the world's oceans through direct discharge of industrial and domestic sewage, burial of toxic and radioactive materials, emergency situations, mainland runoff, shipping, mining, as well as through the atmosphere and with river runoff.

Today's human impact on the marine environment consists mainly of an increase in the rate of pollutants entering the world's oceans and an increase in the type of pollution. Ocean pollution occurs at both the regional and global levels. This process alters the content of various pollutants in the world's oceans.

The oceans and seas cover three-quarters of the Earth's surface, making them natural receptors of pollution from the global circulation of the ocean and atmosphere. The seas have been and continue to be actively used for various types of waste disposal. In the offshore areas of the seas, which are areas of high human activity, the impact of pollution is greater than in the open ocean.

Inland seas such as the Black, Azov, Baltic and Caspian Seas are particularly prone to pollution. The complex nature of the processes that determine the pollution of inland seas makes it very difficult to analyze their ecological status and develop effective methods to combat pollution, develop recommendations on environmental issues.

This edition is dedicated to describing the Black Sea pollution, which is among the most polluted seas on the planet. Two decades ago, many believed that the Black Sea would die in the coming years. However, the marine ecosystem has been preserved.

The book has the following structure. The first six chapters give an overview of the pollution of the seas. The main focus is on chemical pollution, oil and its products, methods of cleaning the sea surface, contamination with plastics. The next five chapters describe the pollution of the Black Sea, the impact on the ecosystems of the Black Sea coastal zone, the self-cleaning properties of the sea and the measures taken to limit pollution. The last two chapters discuss monitoring issues. The first chapter discusses the most general monitoring issues, and the final chapter deals with monitoring methods used in the Black Sea.

Chapter 1 provides general concepts about pollution and contaminants. Natural and anthropogenic sources of pollution in the world's oceans are analyzed. Modern problems of pollution of the World Ocean by the following pollutants and certain types of pollution, among which: oil and oil products; water acidification; runoff—domestic waste discharged into the ocean; thermal pollution; electromagnetic pollution; radioactive pollution; chemical pollution; biological pollution are considered. According to a widely accepted classification, three groups of pollutants are identified: conserved, nutrient and water-soluble. The migration patterns of pollutants and their transformation at the ocean–atmosphere, ocean-bottom sediments and ocean-terrestrial boundaries are considered. The sources and destruction of organic matter in the aquatic environment are considered.

Chapter 2 gives an overview of information on the causes, sources and current levels of microplastics contamination of the world's oceans. By microplastics we mean plastic particles less than 5 mm in size. It has been shown that the longevity of microplastics, combined with the lack of effective recycling facilities, has made marine plastics and microplastics a global problem. The destruction processes of plastics in marine areas are described. It is shown that the main mechanism of microplastics generation is the destruction of larger plastic materials entering the marine environment. The classification of degradation processes of microplastics and conditions of its biodegradation are analyzed. Toxic effect of plastics and especially microplastics on marine environment and biota is discussed. It is shown that the problem of microplastics in the ocean is far from being solved and requires further detailed study.

Chapters 3 and 4 discuss the problem of ocean pollution by oil and oil products and show the dynamics of their distribution depending on the composition of oil and oil products. The main anthropogenic and natural sources of oil pollution in the World Ocean are described. It is shown that the areas of global oil pollution of the seas coincide with the routes of maritime traffic and the estuaries of the largest rivers. Oil degradation processes in the marine environment are analyzed, which include evaporation, emulsification, solubility, oxidation, formation of aggregates, sedimentation and biodegradation, including microbial destruction and assimilation by plankton and benthic organisms. The dynamics of the oil slick and its impact on the characteristics of the boundary layers of the atmosphere and ocean are discussed. Chapter 4 discusses methods for modeling the hydrocarbon slick on the sea surface. Chapter 5 briefly describes the methods and techniques used to clean up oil spills at sea. The possible sources of oil and oil products spills and the likely risks of oil spills during onshore and offshore production, as well as during storage and transportation of oil and oil products due to accidents, are classified. A review of the existing methods and sorbents for liquidating hydrocarbons from the water surface is given, and their characteristics and the principle of operation of the devices currently used for oil spill response are studied. An original method of hydrocarbon spill response using natural sorbent—natural sheep's wool—is proposed.

Chapter 5 briefly addresses marine ecosystems. The functioning of marine ecosystems is analyzed. The classification of the main types of marine ecosystems by depth of location and energy used is described. The spatial organization of energy and substance flows in the marine ecosystem is given. The reasons for different bioproductivity of the World Ocean ecosystems are discussed. Factors affecting the productivity of primary trophic level of marine ecosystems are analyzed. Energy and substance transport by depth and the role of food chains in this process are considered. The biological pollution of the World Ocean is considered, and the influence of biological pollution on the functioning of marine ecosystems is shown. Targeted introduction of "useful" fauna species by humans is considered.

Chapter 6 is devoted to a brief oceanographic description of the Black Sea. The unique features of the Black Sea hydrology are considered, due to its isolation from the World Ocean and its specific connection with it, a large catchment area from which a significant amount of pollution comes, the presence of a pronounced vertical density stratification of waters, leading to a sharp stratification of the sea into a thin desalinated upper layer and deep saline layer. The most important feature of the Black Sea is the presence of a hydrogen sulfide zone, which occupies about 90% of the sea volume. The history of the study of the Black Sea is briefly described. Modern ideas about the specifics of water circulation in the sea are given. It is shown that the main reason for regional differences in climatic changes in the Black Sea from other areas of the World Ocean is the two-layer hydrological structure of waters and the inland position of the sea. The history of the formation of the Black Sea basin is briefly described.

Next Chap. 7 gives an overview of current views on the peculiarities of Black Sea circulation processes and their influence on the formation of the Black Sea hydrochemical regime. The main causes of formation and the main mechanisms of functioning of the sea sulfide zone are considered. The influence of fresh runoff into the sea, 70% of which comes from industrial areas of Europe, is discussed. Estimates of the average monthly values of upward and downward flows of inorganic nitrogen and phosphorus compounds in the deep sea photosynthesis zone of the Black Sea are given, obtained on the basis of averaging the data of long-term observations on the vertical distribution of nitrates and phosphates, as well as calculations of rates of excretion of biogenic elements by plankton organisms. Estimates of flows of mineral forms of nitrogen in the deep sea are given, taking into account their inflow from external sources and exchange processes between layers (surface, deep and near-bottom) of water in the sea.

Chapter 8 deals with the peculiarities of distribution of flora and fauna in the Black Sea taking into account its uniqueness—division into oxygen and hydrogen sulfide zones. The main representatives of the Black Sea flora and fauna are described. Special attention is paid to the marine fish species of the Black Sea with a detailed description of the features of the Black Sea fishing for fish such as the Black Sea Hamsa, Sprat and Mackerel. Current and past catches of fish are briefly presented. The prospects for improving the efficiency of the fisheries and opportunities for aquaculture development in the Black Sea are discussed.

Chapter 9 considers the main sources of pollution in the Black Sea. The data on the Black Sea pollution levels of oil products, heavy metals, household solid waste, plastics and sea radioactive pollution are analyzed. The processes of transport, transformation and disposal of pollutants are considered, and estimates of nutrient inputs to the sea are given, causing eutrophication of shallow coastal waters.

The contribution of the Black Sea states to the Black Sea pollution with oil products is discussed. It is shown that the biggest threat of sea pollution by oil products is the construction of new oil storage facilities and the discharge of oil-containing (ballast) water from ships. Detailed estimates are given of oil pollution in the South Coast of Crimea, Sevastopol Bay and the East Coast of the Black Sea. It is shown that the eutrophication level is related to local discharges of domestic untreated or poorly treated wastewater into the coastal waters of Black Sea towns. The most heavily polluted bays and bays of large cities are Sevastopol, Yalta, Novorossiysk, Gelen-dzhik, Sukhumi, Poti, Batumi and the Pitsunda area.

The information on the Black Sea Track Web (BSTW) operational system for forecasting oil spill spread is given. By the example of the northwestern shelf of the Black Sea, the level of eutrophication of shallow sea areas is estimated, and the causes and consequences of hypoxia and frost zones formation in the studied water area are determined. Modern estimates of the level of contamination of beaches and the coastal zone with solid domestic waste and plastic are given. The level of microplastics pollution of Sevastopol beaches and the East Coast of the sea is analyzed.

The radioactive contamination of the sea is considered in the aspect of the assessment of the consequences for the sea of the world's largest accident at the Chernobyl Nuclear Power Plant (NPP) in 1986.

The nature, genesis and distribution of methane in the Black Sea are considered. It is shown that jet methane gas emissions from the sea bottom are confined to the estuary sections of rivers, shelf edge, continental slope, as well as areas of mud volcanism. Methane flows from bottom sediments to water and from water to the atmosphere have been analyzed. The special environmental role of the process of anaerobic methane oxidation, which is an effective mechanism preventing methane from entering the water column into the atmosphere, is emphasized.

International conventions and agreements of the Black Sea countries aimed at reducing the level of pollution in the Black Sea are briefly outlined.

Chapter 10 explores the self-cleaning capacity of shallow Black Sea ecosystems. The analysis of two methods to assess the self-cleaning capacity of marine ecosystems (balance and synoptic) by calculating the assimilation capacity of shallow marine ecosystems is presented. By the example of individual Black Sea marine areas, the expediency of spatial (zoning) (Dnepro–Bugsky estuary, Sevastopol Bay) and temporal (seasonality) (water area of the Odessa port) detailed assessment of the assimilation capacity of ecosystems in relation to various contaminant complexes is shown.

According to the data of long-term monitoring observations of inorganic nitrogen content in seawaters of the Sevastopol Bay, the balance method assessed the assimilation capacity of ecosystems and E-TRIX trophicity index of the eastern, central, western parts of the bay and the southern part of the bay in relation to nitrites, nitrates and ammonium as priority elements of the biogenic complex in storm and municipal runoff. The example of the central part of the Sevastopol Bay shows the assessment of the assimilation capacity of the marine ecosystem in relation to oil products, as a priority pollutant for areas of developed shipping, performed using the synoptic method.

Chapter 11 discusses the main objectives of the environmental monitoring of the oceans and seas and the problems encountered in their implementation. The main components of the environmental monitoring of the oceans and seas (physical, geochemical and biological) and their goals and objectives are considered. The urgency of organizing effective environmental monitoring of the World Ocean in the production activities of the world's largest companies of the fuel and energy complex on the sea shelf is shown.

The Unified State Information System on the World Ocean (USIMO), developed in Russia, is considered. The objectives of global, national, regional, local, impact monitoring and their correlation are considered in detail.

Information is provided on the distribution of responsibilities among Russian government authorities in implementing various levels of environmental monitoring.

Modern methods and means of ecological monitoring of the environment are considered. The classification of ecological monitoring of environment by sources of influence, factors of influence and studied natural environments is resulted. Modern methods and means of ecological monitoring of World Ocean with an accent on development of technical means of system of complex multilevel ecological monitoring in areas of functioning of sea oil and gas complexes are analyzed.

The final chapter presents an analysis of the structure of environmental monitoring of the Black Sea, its means and methods adopted in the Russian Federation. The standards adopted in Roskom Hydromet of Russia on methods of analysis, schemes of marine sampling stations location, analyzed parameters, peculiarities of formation of the state observation network are described. The differences in European and Russian systems of ecological monitoring of marine environment are considered.

The latest achievements of satellite monitoring of the Black Sea in Russia and prospects for its development are analyzed. The possibilities of space information systems for the calculation of primary production in different areas of the Black Sea are shown on specific examples.

The changes that have taken place in the structure and tasks of environmental monitoring of the Black Sea, adopted in the Black Sea countries after 1992 on the example of Ukraine, are described.

Implementation of the International EMBLAS project "Improvement of Environmental Monitoring in the Black Sea", funded by the United Nations Development Programme (UNDP) and the European Community (EC), is under discussion.

The inclusion in the book of data not only on the pollution of the Black Sea, but also the basic concepts and information about the pollution of the seas and methods of their control, allows using this publication for training students. It is hoped that the book will be useful both for professionals and in the training process.

Contents

Chapter 1
General Concepts of Pollution and Pollutants and Their Nature (Natural and Man-Made Sources of Pollution of the World's Oceans)

1.1 Introduction

Pollution of the environment is considered to be the introduction of new, not peculiar to it physical, chemical, biological components or excess of the natural average level for many years of these components in the environment.

From the environmental point of view, pollution entering the ecosystem changes its functioning significantly. Energy and substance flows, productivity, population size, etc. change. Pollution of the natural environment can come from natural sources (flooding, volcanic eruption, meteorite fall, etc.) and from human activities. Thus, it is necessary to distinguish between natural and anthropogenic pollution. Pollution may have physical, chemical and biological nature.

Let us give a definition of pollution, which was formulated by the UN Expert Group on the Scientific Problem of Global Marine Pollution. "Pollution means the direct or indirect introduction by humans of substances or energy into the marine environment, resulting in harmful effects such as damage to living resources, danger to human health, interference with marine activities, alteration of the useful properties of a water body".

The biological encyclopedic dictionary defines contamination as follows. Pollution of biosphere includes both inflow and accumulation of persistent pollutants, which are almost not destroyed in natural environments (e.g., DDT) and substances having natural mechanisms of decomposition and assimilation (e.g., fertilizers) in amounts exceeding the ability of biosphere to process them and disturbing the natural systems and links in the biosphere that have developed during the long evolution and undermining the ability of natural components to self-regulate.

The first definition is broader, in particular it includes pollution and energy entering the ecosystem.

According to modern concepts [1] pollution is the introduction of new physical, chemical, informational, biological or any other agents, which are usually not

K. Pokazeev et al., *Pollution in the Black Sea*, Springer Oceanography,
https://doi.org/10.1007/978-3-030-61895-7_1

characteristic of the environment; the excess of the natural average level and concentration of agents (within their extreme fluctuations) at a given time leads to negative consequences.

In general, pollution is all that is in the wrong place, at the wrong time and in the wrong amount, which is natural for nature, comes or exceeds the average content level. Pollution brings environmental systems out of equilibrium, is different from the usual observed norms and is undesirable for humans. The agent causing the pollution phenomenon is called a pollutant.

Pollutants are chemical compounds whose concentration in the environment exceeds a certain threshold value, their rate of entry exceeds the rate of their rational assimilation by the respective ecosystems or the rate of their physical, chemical and biological transformation. The presence of contaminants and the growth of their concentrations leads to a disturbance of equilibrium in environmental objects. The consequences of exposure to pollutants or contaminants are called chemical pollution.

When assessing the levels of pollution of a water body or a natural object, the concept of maximum permissible concentrations (MPC) is introduced. Maximum Permissible Concentrations (MPC) is the main value of environmental rationing of harmful chemical compounds in the components of the natural environment. MPC is a norm, maximum quantity of pollutant in environment, which at constant contact or at influence for certain period of time practically doesn't influence on human and animal health and doesn't cause adverse consequences in posterity. The MPC takes into account not only the impact of the pollutant on human health but also its impact on animals, plants, microorganisms and natural communities in general.

1.2 Main Types of Pollution

Physical pollution is associated with changes in physical parameters: thermal, light, electromagnetic, radiation, sound, etc.

Thermal contamination is usually associated with industrial emissions of warm water and various gases. Thermal contamination of water bodies causes their eutrophication and changes the species composition in the body of water. Thermal contamination of the atmosphere can also occur as a result of greenhouse gases entering the atmosphere. Such thermal pollution is secondary in nature.

Electromagnetic pollution occurs as a result of the operation of powerful electrical installations, power lines, radio transmitting devices, including mobile communications. The effect of electromagnetic radiation in its entire spectrum on humans has not yet been studied. The literature describes numerous facts of the negative effect of artificial electromagnetic fields on a person and the ability of a person to feel these fields. For example, some people, being at a distance of ten meters from the power line, are able to determine the moment of its switching on or off. Electromagnetic radiation has negative effects not only on humans. Under power lines, the intensity of electric and magnetic fields is much higher than the background level.

Chemical pollution is the entry into an ecosystem of substances that are quantitatively or qualitatively alien to the ecosystem. Not only does it change the chemical properties of the environment, but ecosystem functioning can be disturbed. Humans supply the environment with compounds that were not previously present in the environment. Therefore there is no natural (natural) way to neutralize them. Examples of chemical pollution are contamination with heavy metals, pesticides, chlorobiphenyls, etc. Negative effects of chemical pollution on metabolism of living organisms are called "ecological traps". As such a trap we can mention the phenomenon of methylmercury accumulation in the human body (Minamata disease—by the name of the area in Japan, where this disease was first discovered). Industrial wastes containing methylmercury were dumped in the Gulf, from where seafood caught by fishermen entered the human body. It took nature more than 40 years to eliminate the effects of toxic waste dumping into the bay. It was only in 1998 local fishermen were allowed to harvest seafood in the bay.

The accumulation of chemicals in the food chain is called concentration in the food chain or bioaccumulation. This phenomenon is particularly dangerous in respect to some biodegradable pesticides and radionuclides, which are used in small concentrations. For example, let us imagine the inflow of DDT into a pond. In the trophic chain, DDT is first transferred to producers (aquatic plants), then to fish (herbivores), then to fish of prey and finally to birds of prey (osprey). When moving from the first to the second trophic level, the concentration of DDT increases 250 times, when moving from the second to the third—another 1200 times, and from the third to the fourth—another 1800 times. DDT has a strong influence on bird populations. The inflow of DDT into the organism of birds violates the formation of eggshells, which impairs the development of chicks and can lead to a reduction or even death of the population. This phenomenon has been well studied, for example, in the California pink pelican population. Similarly, trace amounts of radioactive elements entering water bodies from the nuclear industry are concentrated in the tissues of fish and birds, and are dangerous to humans.

Biological contamination is no less dangerous than chemical contamination. Epidemics of influenza and other diseases are examples of microbiological contamination caused by microorganisms. The spread of pathogenic organisms with wastewater has often been and continues to be the cause of epidemics.

The accidental relocation of animals or plants to ecosystems can cause significant disruption to their functioning. Mnemiopsis ridgewicking with ballast water from the Caribbean into the Black Sea, where it was found to be more resilient than local people in the same trophic chain, caused enormous damage to the marine ecosystem.

The division of contaminants into physical, chemical, biological has certain conditionality and limitations. For example, the thermal contamination of a body of water that occurs when water is used to remove excess heat refers to physical contamination. However, thermal pollution leads to the intensification and restructuring of processes of the biotic component of the ecosystem, chemical transfer processes are disturbed, the species composition changes. Thus, secondary chemical and biological pollution occurs [2].

Another example illustrating the conditionality of division into types of contaminants is given below. Supply of oil products to the surface of a reservoir—chemical pollution. But the distribution of petroleum products on the surface disrupts the processes of gas exchange, evaporation, heat transfer of the pond with the atmosphere. Surface films change the parameters of wind waves, as a consequence there is quenching or weakening of short-wave components of the wave spectrum. Changes in the parameters of surface waves in turn change the structure of the drive layer of the atmosphere and the surface layer of the reservoir—the profiles of wind and current velocity and vertical distribution of pulsations change. The energy and pulse flows between the atmosphere and the reservoir change, i.e. there is a change in the dynamic interaction between the boundary layers of the atmosphere and the reservoir. Recall that the basis of this complex cascade of changes in physical and chemical processes is chemical contamination of the water surface.

The self-cleaning capacity of the natural environment depends on the characteristics of the pollutant and the biocoenosis. In case of pollution, for example, of water bodies, pollutants can be divided into the following groups [3]. The first group includes conserved pollutants that do not decompose or decompose very slowly in the natural environment. Reduction in the concentration of conservative pollutants is due to the processes of dilution, mass transfer, sorption, bioaccumulation, and others. Self-cleaning in this case has an apparent character, as there is only redistribution and dispersion of conservative pollutants in the environment. Local pollution is reduced by expanding the pollution area. The total amount of pollutants does not change. The second group includes biogenic pollutants that participate in the biological cycle. The self-cleaning of the natural environment is done through biochemical processes.

The third group may include water-soluble substances that do not participate in the biological cycle. Self-purification of these pollutants is done through chemical and microbiological transformation.

The response of the ecosystem to the introduction of different types of pollutants into the ecosystem will vary. With an increase in the concentration of pollutants involved in the biological cycle, there is an initial increase in the bio-productivity of the ecosystem, with a further increase in the concentration of pollutants there is a decrease in productivity and possible loss of ecosystem life.

When contaminants that do not have natural ways of utilization enter the ecosystem, as the concentration of the contaminant increases, the bioproductivity of the ecosystem reduces.

Anthropogenic pollution is usually of a local nature. The distribution over a larger area is due to the multiplicity of migration routes.

Pollutants entering the atmosphere are involved in global atmospheric circulation. Deposition of atmospheric pollutants on the surface of water bodies leads to their contamination. Deposition of atmospheric pollutants on the surface of plants causes contamination of biota (extra root inflow of pollutants).

Contaminants entering a water body participate in the circulation of the water mass. The evaporation of water can cause atmospheric pollution, the transition of polluted water into soil or biota.

In a similar vein, pollutants that originally entered the biota or land surface cause air and hydrosphere pollution can be listed. A cursory look at the processes of pollutant migration shows that physical processes of mass transfer play a crucial role in the distribution of pollutants. The role of physical mechanisms at the interface and in the boundary layers is particularly important.

The multiplicity of pollutant migration routes leads to the need for an integrated approach to studying this process. It is impossible to solve the task of pollution transfer only within the framework of physical, chemical or biological processes study.

Most of the pollution comes to the seas from land (40–45%). Not only does chemical pollution pose a major threat to marine ecosystems, but large quantities of nutrients are released into the sea from the river runoff, causing eutrophication of seawater. Atmospheric transport is the second most important source of pollution, roughly equal to the share of pollution from river runoff. The third place (approximately 20%) is occupied by local sources of pollution, including municipal runoff, dumping (dumping of pollution), flushing of industrial waste, etc.

Table 1.1 presents anthropogenic and natural sources of pollutants entering the World Ocean [1].

Many processes link the oceans to other areas of the natural environment. These are the interaction of the ocean and atmosphere, the passage of biogeochemical cycles of the most important chemical elements that determine the circulation of matter and energy in natural ecosystems, the powerful photosynthetic activity of algae, regulating the balance of oxygen and carbon dioxide and other phenomena of global character. V.I. Vernadsky wrote "The world ocean with its complex equilibrium processes is not isolated in the Earth's crust. Its substance is in close exchange with the atmosphere and land, and this exchange is of great importance not only for the chemistry of the sea, but also for the chemistry of the entire earth's crust". However, at present, anthropogenic activity has a significant impact on the supply of many chemical compounds to the marine environment. Recent studies have shown that the anthropogenic component of the runoff of a number of pollutants into the world's oceans (lead, oil, mercury, arsenic) is comparable to or even greater than the natural component (Table 1.2).

In the coastal zones of the seas, increasing anthropogenic impacts have resulted in progressive eutrophication and microbiological contamination of seawater and hydrobionts. At the same time, chemical concentrations are rapidly increasing to critical levels.

In open areas of the seas, hydrobionts and ecosystems as a whole are beginning to be affected by low intensity factors—low doses of resistant chemical compounds, the danger of which is chronic exposure. Both intensive and low-intensity pollution pose a threat to the ecological well-being of the hydrosphere as a whole.

Pollutants entering the hydrosphere in general and the oceans in particular are distributed unevenly, contributing to increased pollution of coastal areas, the

Table 1.1 Anthropogenic and natural pollutants entering the world's oceans

Type of contamination	Sources	
	Natural	Anthropogenic
Petroleum hydrocarbons	Outlets of oil and gas; river and terrigenous runoff, volcanoes, bacteria in the water column, atmosphere	Asphalt roads, transport, mining, aerosols
Suspended substances	River and terrigenous runoff, current-induced disturbances, high biological productivity, atmosphere	Agriculture, fishery (trawling), dredging (in ports, rivers, canal-lahs), industrial and domestic sewage, drilling
Heavy metals	Volcanoes, river and terrigenous runoff, cracks, faults in the Earth's crust, bottom sediments, decomposition of organisms	Industrial and domestic sewage
Radioactive materials	River and terrigenous runoff, volcanoes, cracks, faults in the Earth's crust, deposits, atmosphere	Industrial and domestic sewage, nuclear power plants, nuclear weapons testing
Biogenic substances	River and terrigenous runoff, sedimentation, turbidity, biological cycles, atmosphere	Domestic sewage, agriculture, liquid clay
Thermal effects	Volcanoes, cracks, crust faults, overheated tropical lagoons, estuaries	Emissions from refrigeration plants, use of ocean thermal energy
Rapa	Salt lenses, cracks, earth's crust faults, shallow lagoons	Industrial emissions, including the intake of brine from salt storage tanks
Biological oxygen consumption	"Red tides", eutrophication, oxidation and transformation	Domestic and industrial emissions, canning industry waste

Table 1.2 Anthropogenic load (t/a) on the world's oceans by major pollutants as compared to natural flows

Pollutant substance	Stoke		Share of anthropogenic runoff (%)	Flow into the ocean	
	Natural	Anthropogenic		Stoke onshore	Atmospheric deposition
Lead	1.8×10^5	2.1×10^6	92	$(1–20) \times 10^5$	$(2–20) \times 10^5$
Mercury	3.0×10^3	7.0×10^3	70	$(5–8) \times 10^3$	$(2–5) \times 10^3$
Cadmium	1.7×10^4	1.7×10^4	50	$(1–20) \times 10^3$	$(0.5–14) \times 10^3$
Petroleum	6.0×10^5	4.4×10^6	88	$(3, 4) \times 10^6$	$(3–5) \times 10^5$
Chlorinated hydrocarbons:					
PCB	–	8.0×10^3	100	$(1–3) \times 10^3$	$(5–7) \times 10^3$
Pesticides, dibenzodoxins, dibenzofurans	–	1.1×10^4	100	$(4–6) \times 10^3$	$(3–7) \times 10^3$

euphotic layer and the hydrofronts zones where most organic matter is concentrated. Contaminants accumulate and primarily affect the ecosystems of the outer contours of the sea and the interfaces, the so-called critical zones, where the most active geochemical processes take place and where abundant and diverse forms of marine communities are developed. Thus, the water-atmosphere interface serves as a habitat for the assemblage of neuston and pleiston forming organisms. Neuston organisms live in the surface film, so they are most exposed to oil products and well soluble in them polychlorinated compounds, pesticides. Most of the contaminants pass through the water column and can be deposited on the bottom as part of suspension. Toxicants can thus accumulate in the bottom sediments and in certain situations, the bottom sediments can become a source of "secondary" contamination.

In recent years, the major role of atmospheric transport in ocean pollution has been recognized. Increased concentrations of pollutants have been detected in surface waters and especially in the surface micro-ocean. Organic natural and oil films affect the physical and chemical properties of the ocean-atmosphere interface. They inhibit the processes of gas exchange, quench capillary waves, change the temperature of the surface micro-layer and the nature of the processes of bubble formation, which contribute to the natural self-cleaning of water masses.

The ocean and atmosphere form a single "thermodynamic machine", whose activities affect the climate system. Disturbance in the intensity and nature of the circulation of gases and substances between the ocean and the atmosphere causes climate change, which is specific to the polar and equatorial regions of the Earth.

Thus, the surface micro-ecosystem, as a critical zone of accumulation of pollutants in the world's oceans, is subject to study to clarify biogeochemical cycles and balance of elements, as well as to forecast the consequences of anthropogenic impacts on open ocean systems and the Earth's climate.

The ocean-to-land contact zone is a chemical-geochemical barrier between mainland runoff and the open ocean. It is the area with the highest biological activity in the world's oceans. The coastal areas, which cover 13% of the total ocean area, produce 40% of all primary organic matter in the marine environment. It is important to note that the coastal areas and upwelling zones are home to more than 90% of the world's oceanic fishery resources.

The coastal areas are the areas most actively exploited by humans and provide the greatest economic benefits. Consequently, it is the area with the greatest human impact on the world's oceans. Currently, about 50% of the world's population lives near the ocean or is directly linked to its coastal zone. About 50% of the world's largest cities, with populations exceeding 1 million, are located near estuaries or bays of the oceans.

In addition to the pollution of river waters reaching the coastal and open areas of the world ocean, in addition to direct discharges of domestic, industrial and agricultural runoff, human beings affect the marine ecosystems of coastal areas in the course of port dredging, construction of ports and piers, the active withdrawal of biological resources, the laying of underwater pipelines and other underwater work. All these impacts, as well as inland run-off, including river run-off, inevitably lead to changes in the physical, chemical and biological processes in the ocean-to-land contact zone.

River runoff is the most important way for contaminants to enter the marine environment. Estimating the amount of pollutants carried out by river water is crucial to studying the assimilation capacity of coastal ecosystems and to studying the biogeochemical cycles of chemical compounds. With river discharge, 20–109 tons of suspended solids and dissolved salts, including metals and organic pollutants in dissolved form, as well as in association with suspended solids, enter the oceans. On boundary fronts in estuaries and on the shelf, they interact with water and open ocean suspended matter. A certain proportion of dissolved and suspended contaminants are discharged into open waters and into deep oceans. The concentration of suspended solids in coastal waters is higher than in the open ocean. This phenomenon is one of the most important problems in the geochemistry and microbiology of the ocean-land boundary zone.

The chemical composition of inland waters, which are influenced by humans, depends on precipitation, temperature, soil type and mineral composition of the bottom rocks of the reservoir, morphology, flora and fauna, as well as the duration of exposure to these factors. Pollutants can enter the hydrosphere with atmospheric precipitation, soil water and river runoff, as well as with domestic and industrial runoff and drainage flows from agricultural activities. Acidification of fresh water occurs when the rate of substitution of soil cations by hydrogen (H+) exceeds the rate of cation intake due to weathering. Freshwater acidification is particularly noticeable in mountainous areas with high rainfall (and therefore high acid flow) on steep slopes (where the result is short water retention times in the soil) and slow processes of erosion and cation supply.

The critical boundary zone water—bottom sediments—is the final link of transformations and burial of many chemical compounds spread in the biosphere. This zone is the area of both accumulation and transformation of chemical substances, including pollutants, which circulate in the water column, enter the ocean from land, from the atmosphere and reach the ocean bed with suspended matter or with descending water flows. This zone is also the critical zone where the concentrations of pollutants reach their highest values.

Indeed, contaminants that are common in the ocean's water masses are either destroyed there or somehow reach the bottom layers. Final disposal of sediment occurs after it has passed through the water-bottom boundary zone. It is here that organic detritus, calcium, silicon shell remains and other chemically active materials undergo various transformations.

Despite the remoteness of the ocean floor, especially in open areas, high concentrations of high-molecular organic matter and heavy metals are increasingly being found in the upper layers of the soil from areas of anthropogenic activity. Bottom sediment pollution has progressed particularly in the last 30–40 years and is most characteristic of inland seas and coastal ocean areas.

The water-bottom surface and, above all, the bottom sediments themselves, represent a permanent source of secondary pollution of seawater, which occurs during silt sedimentation due to hydrological phenomena or bottom animal activities (biological turbulence).

Characterizing the most important natural processes taking place at the water-bottom sediments border and determining the elimination of pollutants or their return to the water column (secondary pollution), it is impossible not to mention biological disturbances, which are now considered as a global ocean phenomenon.

So, the critical zone of water-bottom sediments is a dynamic medium, where the speed of transport and the speed of reaction influence the conditions much more than the chemical equilibrium of processes. Many chemical compounds that reach the ocean floor then return to the water column and are redistributed there.

Various processes take part in the transformation of pollutants in marine ecosystems: physical–transfer, adsorption, dilution, concentration at evaporation, proceeding without changing the composition of pollutants; physical-chemicaldecomposition, redox breaks of chemical chemical oxidation, decomposition, redox breaks of chemical bonds, chains, formation of new compounds as a result of interaction of pollutants with hydrosphere components; as a result of hydrolysis reactions, complex formation, substitution with changes in the composition of pollutants. And, finally, biological processes related to the inclusion of a contaminant in the trophic cycle, causing a chronic or acute lesion of the biocoenosis, and completing the cycle of the contaminant either by returning to the ecosystem of its constituent elements in the case of a fully decomposing substance or their final burial in bottom sediments in the case of non-degradable substances [4]. It is also possible to remove the pollutant outside the ecosystem under consideration as a result of the current flow system or wind activity.

Let us focus in more detail on the physical processes of pollutant transport in the world's oceans. Because of the great dependence of elements of the ocean ecosystem on large-scale circulation, local pollution and its negative effects in the world's oceans become global.

Vertical mixing and ordered horizontal and vertical movements carry toxic substances over long distances and into the deep ocean. Small-scale mixing of water masses, while not conducive to long-distance transport, is of significant environmental importance in terms of changing properties and local concentrations of pollutants. Even when contaminants arrive uniformly at the surface of different areas of the world's oceans, hydrological factors (diffusion, advection) predetermine the "spotty" distribution of contaminants.

The relationship between physical, chemical and biological processes is particularly close in the areas of frontal zones, upwelling and downstream.

The functioning of the ocean's near-surface biotope system is largely determined by surface water mixing, which is caused by wind force and speed and by the processes of heat exchange between the ocean and the atmosphere. Consequently, meteorological conditions have a very direct impact on the surface layer of the ocean, where pollutants are concentrated and neuston biocoenoses develop, which play an important role in the functioning of the biotic component of the ocean and in the destruction of organic compounds in its surface films.

Thus, the dynamism of the ocean environment, a set of constantly operating critical physical phenomena determine the basic laws of distribution of pollutants in the world ocean, namely:

Long-range transport of pollutants by intensive currents to open areas of the ocean, its oligotrophic zones, northern ecosystems, upwelling areas, coral reefs and others.

Concentration of pollutants in areas of convergence of heterogeneous water masses, structural currents, in estuaries and zones of quasi-stationary cycles.

The transport of contaminants into bottom horizons, deeper layers of the ocean and their accumulation in marine organisms, suspended matter and sediments.

Accumulation of pollutants on the surfaces of the ocean-atmosphere, ocean-land, water-bottom sediments.

There is now a growing awareness of the need to assess transoceanic transport of contaminants and determine its impact on oligotrophic and productive ecosystems in the open ocean.

In the transformation of pollutants in water bodies, redox processes are of great importance. The pH and Eh acidity of the redox potential can determine the behaviour of elements and their compounds in the environment. Under highly oxidized conditions Eh 0.6–1.2 V water is broken down into ions O and H_2, and under reduced conditions Eh 0.0–0.6 water is restored to hydrogen. Diagram Eh—pH gives an opportunity to estimate the state of this or that element under changes in acidity and redox conditions. Depending on Eh and pH different forms of elements can be formed, for example Mn and Fe. Manganese and iron, potential electron acceptors under oxidizing conditions occur as insoluble oxides Fe(III) and Mn(IV). Under reducing conditions at Eh values from zero and below towards negative values these oxides can dissolve to form soluble Fe^{2+} and Mn^{2+}. By the way, iron is soluble only at low Eh and in acidic environment. The same transformations depending on Eh medium are typical for sulphur compounds. In an oxidizing environment prevails sulfur of maximum state +6, in reducing conditions of sulfur exists in the form of S^{2-} negative ion.

1.3 Ocean Acidification as a Result of Pollution

The burning of fossil fuels not only pollutes the atmosphere, but also the ocean. The oceans absorb up to a quarter of all anthropogenic carbon emissions, altering the pH of surface waters and oxidizing the sea.

The problem is worsening—the oceans are now oxidizing faster than in about 300 million years. It is estimated that by the end of this century, if we keep up with our current rate of emission, the surface waters of the ocean could become almost 150% more acidic than they are now.

What happens when biochemical processes in the ocean are disrupted?—There are changes in marine ecosystems and coastal economies that depend on them.

First we'll take the reefs and the clams. To build their shells and skeletons, creatures such as mussels, clams, corals and oysters require calcium carbonate (the same compound as in chalk and limestone). But the level of carbonate in the ocean drops when acidity rises, threatening the survival of these animals. Bivalve molluscs are at the beginning of the food chain, so increases in ocean acidification have a negative impact on fish, seabirds and mammals. More acidic waters also contribute to the bleaching of coral reefs and make it difficult for some fish species to recognize predators and for others to hunt for prey.

Organic matter in the waters of seas and oceans Organic matter is the main source of energy for processes occurring in the water column and bottom sediments.

If we take from Horn [5], that the composition of organic matter can be roughly expressed by the formula $(CH_2O)_{106}\,(NH_3)_{16}\,H_3PO_4$, then in normally aerated sea water oxidation of organic matter is carried out by oxygen and it can be represented by an equation:

$$(CH_2O)_{106}\,(NH_3)_{16}\,H_3PO_4 + 128O_2 = 106\,CO_2 + 122\,H_2O + 16\,HNO_3 + H_3PO_4 \quad (1.1)$$

If there is a lack of oxygen in the system, this process can be represented by an equation:

$$(CH_2O)_{106}\,(NH_3)_{16}\,H_3PO_4 + 84.8\,O_2 = 106\,CO_2 + 42.3N_2 \uparrow + 148.4\,H_2O + 16\,HNO_3 + H_3PO_4 \quad (1.2)$$

In addition to the reaction (1.2), ammonium oxidation may occur:

$$5NH_3 + 3\,HNO_3 = 4\,N_2 \uparrow + 9\,H_2O \quad (1.3)$$

After the amount of oxygen drops to 0.11 ml/l and the stock of NO_2 and NO_3 is exhausted, the decomposition of organic matter occurs due to oxygen sulfates from bacterial reduction sulfate:

$$(CH_2O)_{106}\,(NH_3)_{16}\,H_3PO_4 + 53(SO_4^{2-}) = 106CO_2 + 53\,S^{2-} + 16NH_3 + 106H_2O + H_3PO_4 \quad (1.4)$$

The main sources of organic carbon are plankton and macrophyte detritus. Organic matter of natural origin can be present in seawater, either as dissolved organic matter or as suspended organic matter.

Dissolved organic matter (DOM) in seawater is represented by soluble amino acids, polysaccharides and lipids. It is produced in seawater mainly by phytoplankton, with the contribution of the primary production of the DOM equivalent to that of land sources. In rivers, approximately half of the dissolved organic matter is in the form of highly oxidised depleted (C/N = 15–85) soluble humus substances or nitrogen-enriched organic substances associated with mineral sediments. A specific feature of the destruction of organic matter in coastal zones is the effect of suspended material. In offshore zones, the concentration of suspended solids increases several times. Much of the suspended material, including organic material, is brought into coastal zones with river runoff. This suspended material contributes to the rate of destruction of organic matter. On the shelf, softened bottom sediments have a great influence on the rate of gross sedimentation of suspended organic carbon and organic nitrogen, amounting on average to 80–90% of the total Corg and Norg suspended solids deposited on the shelf [6]. An important feature of organic matter destruction in coastal zones is also the influence of biological processes. The efficiency of these processes depends on the form of organic substrate and the ability of the enzymatic system to interact with organic compounds. The combination of organic matter with metal pollutants and their sorption on the surfaces of colloids of different nature can "protect" organic matter from the effects of enzymes and thus slow down its destruction. The rate of destruction of organic matter depends to some extent on the level of trophicity of the shallow water area.

It is known that starting from a certain level of trophicity, the source of secondary eutrophication is bottom sediments. Three groups of aquatic ecosystems are distinguished depending on the values of primary production and rate of organic matter accumulation in bottom sediments [7].

I accumulation of organic substance of plankton origin (reservoirs with undisturbed ecosystem) less than 20%;

II accumulation from 20 to 50% (reservoirs where the share of macrophytes is 25%);

III accumulation above 50% (share of macrophytes and allochton detritus in organic matter of bottom sediments >70%).

In oligotrophic (70 gC/m^2 per year) and mesotrophic (150 gC/m^2 per year) waters, more than 50% of plankton detritus decomposes, in eutrophic waters (300 gC/m^2 per year)—this value decreases to 20–40%, i.e. the bulk of detritus decomposes in bottom sediments. Similarly to the organic matter, the rate of accumulation of nitrogen and phosphorus in the bottom sediments of reservoirs has been established, depending on their trophicity. More than 90% of the nitrogen in bottom sediments is organic compounds, of which more than 45% is amino acids. The cycling of elements in the coastal ecosystem is influenced by high solubility of mineral nitrogen compounds and low solubility of phosphorus compounds. And the presence of streams of nitrogen and phosphorus from the bottom in a eutrophized body of water is determined (all other conditions being equal) by the growth of organic matter content and the intensity of its destruction in sediments.

The decomposition rate of dissolved organic matter in river mouths depends on the salinity of the water and the time of DOM residence in the downstream zone. Typically, the components of the decomposition rate of dissolved organic carbon

(DOC) indicate that the DOC content in low salinity water decreases by 5% and by 0.8% in higher salinity water during the residence time (max. 4 days) of river estuarine water. It was found that the bacterial decomposition of the DOC is negligible during such a short period. If the period of stay in the estuary zone is extended to 80 days, 20% of the river DOC will be decomposed, and the remaining 80% will be removed from the estuary. Thus, decomposition rates in estuarine areas are relatively low.

Based on a study of the behaviour of DOC in estuaries, it has been concluded that riverine delivery of DOC to the sea can be as high as 50% of the high seas DOC. In shallower areas, the transformation of suspended organic matter (SOM) takes place in the water column and at the sediment-water boundary. SOM at the bottom is a food source for benthic fauna and heterotrophs. Net suspended organic carbon deposition (SOC) is 10% of the main SOC. Of the deposited SOM, 50% is phytoplankton carbon. It is shown that 88% of suspended amino acids are decomposed or transformed into soluble forms during deposition and 12% are deposited at the bottom.

Proceeding from the general idea of the behaviour of natural organic matter in shallow seas and oceans and using the observation data obtained for the north-western shelf of the Black Sea, the work proposes a scheme of the cycle of organic matter of natural origin for the northwestern shelf of the Black Sea [8]. It is shown that the amount of river suspended organic matter is a significant part of the total content of sea water, and it arrives unevenly, about 50%—during the spring. The north-western shelf of the Black Sea differs from other areas of the world ocean in that the proportion of soluble organic matter supplied by rivers is relatively small in relation to its content in seawater. It has been established that approximately 50–70% of primary production is decomposed in the water column and carried away by currents. Approximately 80% of suspended amino acids are decomposed or transformed into a soluble phase.

Quantitative characteristics of microbiological oxidation of organic matter are BOD and COD values. BOD is an index used to characterize the degree of sewage pollution by organic impurities capable of decomposition by microorganisms with oxygen consumption.

Radioactive pollution of the world's oceans Radioactivity—transformation of atomic nuclei of some chemical elements into nuclei of other elements. Spontaneous transformation of atomic nuclei is called natural radioactivity.

In contrast to natural radioactivity, artificial radioactivity is the radioactivity of isotopes derived from nuclear reactions. Nuclear reactions are artificial transformations of atomic nuclei that are caused by their interaction with different particles or with each other.

Radioactive and stable isotopes are of interest from the point of view of meteorology and climatology, as their concentration in natural objects and, in particular, in the atmosphere, serve as a kind of chronometer. Such a chronometer makes it possible to estimate the time of stay of the isotope in the object under consideration. Geochronological assessments help to build a picture of the dynamics of natural objects in historical terms, spreading our knowledge of the Earth's development thousands and millions of years ago beyond the time period of quantitative measurements.

Under conditions of natural radioactivity, the nuclei of radioactive isotopes of one chemical element turn into isotope nuclei of other chemical elements. In this case, certain particles are emitted: α, β-radiation, electromagnetic γ-radiation. α-radiation is a flow of helium nuclei, β-particles are a flow of fast electrons, γ-radiation is a hard electromagnetic radiation with the highest penetration capacity as compared to other radioactive rays.

Among the heavy nuclides there are three families characterized by natural radioactivity: the family of uranium ($^{238}_{92}U$), the family of thorium ($^{232}_{92}Th$) and the family of actinia ($^{231}_{92}Ac$). In each natural radioactive series, the chain of radioactive transformations ends with stable nuclei of lead isotopes: the nucleus $^{206}_{82}Pb$ for the uranium family, $^{208}_{82}Pb$—for the thorium family, and $^{207}_{82}Pb$—for the actinium family.

Artificial radioactivity of isotopes is related to the instability (stability) of the atomic nucleus, which as a result of a nuclear reaction can be "overloaded" with neutrons or protons.

While natural radioisotopes constantly enter and leave the atmosphere in the natural cycle, the intensity of artificial radioisotopes entering the atmosphere depends on anthropogenic activities and the use of processes based on nuclear reactions.

Radioisotopes present in seawater are divided into three classes:

Natural radionuclides with very long half-lives that have existed since the planet's formation, as well as their daughter radionuclides with shorter half-lives that are continuously renewed by the decay of the mother element.

Natural, relatively short lived nuclides formed by processes such as cosmic radiation in the atmosphere.

Artificial nuclides arising from the contamination of the planet by human activity.

The study of radioactivity in the waters of the World Ocean began after World War II. These studies were stimulated by the intensive development of nuclear power, explosions of atomic and hydrogen bombs. In 1954–1955 Japanese scientists in the area of Bikini Atoll traced the path of water masses contaminated with radioactivity due to atomic bomb testing in the atmosphere. The products of the explosions together with the polluted oceanic waters in 4 months traveled a distance of 2000 km, and for 8 months—7000 km. It is clear that all inhabitants of the planet are at risk.

With the development of the nuclear industry, the issue of radioactive waste burial also arose. In the 50s of the last century, the Black Sea was considered to be an ideal disposal site for such waste. At that time it was believed that the deep layers of the Black Sea were calm and the age of the bottom waters was $\sim$1000 years, i.e. there was almost no exchange with the surface layers. To clarify this issue, research into the dynamics of Black Sea waters was intensified and it was shown that the age of the deep water masses in the Black Sea is only $\sim$130 years. The Department of Sea and Land Water Physics of the Faculty of Physics also took part in these works. The staff of the Department of Sea and Terrestrial Physics made instrumental measurements of flow velocities in the bottom layer of the Black Sea, which showed that the velocity at the bottom in some areas reaches the value of several tens of cm/s. The natural measurements of turbulent characteristics near the bottom allowed us to determine the true values of exchange coefficients in the bottom sea layer. As a result of studies carried out, the project on radioactive waste disposal in the Black Sea was closed at that time. However, the problem of decontamination and disposal of radioactive waste has become more and more urgent over the years and requires its resolution. At present, as a rule, radioactive waste disposal in the sea is performed under international control.

The study of ocean water radioactivity is not only of purely practical importance. It is one of the methods to study the processes occurring in the oceans.

The main tasks of modern nuclear hydrophysics are to identify sources of isotopes in the ocean, to study the fields of radioactivity on the surface of the ocean, to study the deep distribution of isotopes, to create methods for predicting the radiation situation in the ocean, to develop research methods and methods for measuring the radioactive background of the world ocean.

Surface contamination At present, almost 20% of the world's oceans are covered by organic films. This includes substances of anthropogenic and biogenic origin. Anthropogenic pollution is especially high in areas of human economic interest—shelf, river estuaries, large shipping routes. Oil and oil products form the basis of this type of pollution.

In the presentation of this subsection, we follow [9]. Besides oil products, synthetic surfactants (SPAVs) are found in coastal waters. A large number of detergents used as detergents, stabilizers of foams and emulsions, evaporation retarders, etc. are used in industry and household. When waste water enters the ocean, it also participates in the formation of surfactant films. These substances are diverse in composition and include alkyl sulphates and sulfonates, alkylbenzene sulfonates (with a chain length of 12–16 carbon atoms), polyethers of oxyethylated alcohols, etc. Basically all these substances are well soluble in water.

Another source of anthropogenic pollution entering the ocean is adsorption from the atmosphere, from so-called aeolian suspension—dust particles carried by the wind. For samples taken in the Pacific Ocean, one cubic metre of air contains between 0.1 and 0.8 μg of organic matter. These are carbonic acid derivatives with a chain length of 12 to 18 carbon atoms. Moreover, even bases prevail over odd ones, and the maximum is $C_{16:0}$ carbon (palmitic acid). Unsaturated hydrocarbon chains have also been found, but their content is lower compared to saturated ones. For them, the maximum content corresponds to group $C_{18:1}$ (oleic acid).

The formation of natural surfactants can occur biogenically through the life activities of zooplankton, phytoplankton and other marine organisms and plants. Despite the fact that bioproductivity is maximum in the sea water column, suspended and dissolved biogenic surfactants are carried to the surface by air bubbles. The material of bubble walls saturated with adsorbed surfactants is transferred to surface films.

Detailed chemical analyses of biogenic organic surfactants have not yet been conducted. Although it has long been known that the physical properties of natural films from different samples are similar. This can be seen from surface pressure dependencies, which, indirectly, may confirm the identity of their composition. On the other hand, the physical properties of typical samples of natural films differ from the properties of their constituent compounds. This may occur due to the fact that natural films are a complex mixture of a wide range of surfactants. Besides, natural films often represent a layer of the order of several dozens of molecular sizes.

It is known from the results of analysis of samples of natural surfactant films that they are complex complexes of the main classes of organic compounds: lipids, proteins, carbohydrates, etc. (for example, sterols, humic substances). An approximate distribution between the classes was found: proteins ~50%, carbohydrates ~30%, lipids ~12%. The breakdown by percentage composition is conditional, because in the natural film organic substances present not in pure form, but in a combined state. It is noted that even a small addition of lipids can have a significant effect on the surface pressure in the film. A comparison of natural films with artificially composed known components has shown that the most stable components are fatty acids and their esters, and to a lesser extent proteins, carbohydrates and hydrocarbons.

The predominance of saturated carboxylic acids in natural films indicates that chemically stable surfactants are accumulated at the interface surface. Despite the fact that they are contained in marine organisms in lower amounts than unsaturated acids, their high resistance to microbiological and photochemical effects leads to the observed balance. Note that due to their origin, surfactant biogenic films are characterized by seasonal variations in composition.

Along with films, a special danger for water bodies is posed by pollution concentrated in surface microorganisms [10]. The concentration of contaminants in microorganisms is more than an order of magnitude higher than in the underlying layers, which can lead to harmful effects on marine ecosystems. The increased concentration of oil pollution is also observed at the water-suspension interface located in the bottom part of the reservoir.

Surface contamination occurs in the form of slicks—areas of smoothed water surface. Contaminations reduce the high-frequency components of the wind wave, the reduction of high-frequency components of the spectrum of surface waves is manifested in the smoothing of the water surface. Surface tension in natural films (in the absence of visible slick) is reduced by relatively pure water by 0.5–1.5 din/cm, and for visible slick by 6–9 din/cm. In some cases of high contamination, lower surface tension coefficient values are also observed. Thus, for the New York Bay water area, a drop to 51 din/cm was observed, and in some areas of the Atlantic surface tension coefficient values of 65 din/cm were observed [11]. A significant decrease in surface tension has been observed in coastal areas and areas of intense shipping traffic. This decrease correlates well with the concentration of synthetic surfactants in the surface layer. Natural films mainly exist at surface pressures below critical. Under natural conditions they are able to survive in the wind up to 5–7 m/s. At high wind speeds the film is destroyed and can be restored at the termination of the dispersing factor, depending on the type of influence and composition of the film, in 10 s–10 min. Taking into account the fact that the average wind speed in the ocean is 6–8 m/s, it seems that surfactant films in the marine environment exist in a dispersed state, naturally excluding areas of anthropogenic pollution, high bio-productivity and conditions close to calm. In this state, their elasticity is insufficient to have a quenching effect on waves and slick formation. However, the formation of tightly packed films and slicks is possible in areas where currents converge (e.g. through internal waves or Langmuir circulation) and wind surges.

1.4 Conclusions

The first section deals with the current problems of pollution of the world's oceans with the following pollutants: Oil and petroleum products; Acidification of water; runoff—domestic waste discharged into the ocean; Thermal pollution; Radioactive pollution; Biological pollution.

The above information shows that rapidly developing technological progress has initiated environmental problems on the planet, including pollution of the world's oceans.

It is concluded that only environmental scientists speak about the urgency of the problem, and society and political forces are not yet sufficiently involved in its solution, although it is possible to stop the negative impact of man on the environment only through joint efforts.

References

1. Israel Yu A, Tsyban AV (1989) Anthropogenic ecology of the ocean. L. Gidrometeoizdat, 528 p
2. Ivanov VA, Pokazeev KV, Sovga EE (2006) World ocean pollution, Moscow State University, 163 p
3. Scurlatov YI, Dooka HG, Misiti A (1994) Introduction to ecological chemistry pM.: Higher Shk., 400 p
4. Sovga EE (2005) Pollutants and their properties in the natural environment. Sevastopol, Ecosi-hydrophysics, 237 p
5. Horn R (1972) Marine chemistry (water structure and hydrosphere chemistry) M. Izd., 399 p
6. Sovga EE, Zhorov VA, Boguslavsky SG (1998) Basic laws of the processes of sedimentation of suspended organic substance in shallow zones of the seas of moderate latitudes. Marine Hydrophys J 6:60–64
7. Martynova MV (1998) Law of processes of accumulation, transformation and extraction of nitrogen and phosphorus compounds from the bottom of reservoirs: abstract of dissertation. Doctoral thesis, Rostov-on-Don, 35 p
8. Lisitsyn AP, Koronovsky NV, Shreider AA, Glazyrin EA, Shestopalov VL, Yanina TA (2018) The black sea system. Scientific World, Moscow, 808 p

9. Lazarev AA, Pokazeev KV, Shelkovnikov NK (1998) Physical and chemical uniformity of the ocean surface and surface waves. part 1, physics and chemistry. Moscow State University, 134 p
10. Mikhailov VI (1992) Surface microscope of the world ocean (hydrochemical and physical features). St. Petersburg, Gidrometeoizdat, 276 p
11. Nesterova MP (ed) (1989) Methods and tools to combat oil pollution in the world's oceans. Problems of chemical pollution of the world ocean waters. L. Gidrometeoizdat, 206 p

Chapter 2
Current Problems of the World Ocean Pollution by Microplastic

2.1 Introduction

The widespread use of plastic products in industry and household has led to the problem of accumulation of associated waste. Since the mid-twentieth century, there has been an annual increase in demand for plastic products, which now stands at about 300 million tons, with 2/3 of plastic products—packaging materials and disposable items. One of the main reasons is the low cost of polymers, their low weight, bioinert, durability and resistance to wear.

The following plastics are used to manufacture these products: polyethylene (PE), polypropylene (PP), polystyrene (PS), polyethylene terephthalate (PET) and polyvinyl chloride (PVC) [1]. All these compounds are represented in the debris structure in the coastal-sea zone. According to statistics, more than 80% of the sources of marine pollution are land (coastal), including waste from recreational activities. It is established that at least 60% of marine debris is plastic. About 18% of plastic waste is from fisheries where polyolefins (PE and PP) and nylons are actively used to make gear Mariculture may also be a source of plastic waste in the ocean [2].

The study of global ocean pollution by plastic waste is one of the actively developing areas in the anthropogenic ecology of the ocean.

Longevity is a common feature of most plastics, and it is this property, combined with a lack of capacity to effectively dispose of and recycle plastic with a finite life, that has made marine plastics and microplastics (plastic particles smaller than 5 mm) a global problem, both transboundary and social, economic and environmental [3].

A frightening feature of microplastics is its rapid and widespread spread in the world's oceans. The "plastic era" began in human life only in the middle of twentieth century, less than 70 years ago, and today microplastics has already been found in the bottom sediments of the ocean hollows, and in the waters of the Antarctic, and in the ice of the Arctic (up to 1 million fibers per cubic meter of ice!),

K. Pokazeev et al., *Pollution in the Black Sea*, Springer Oceanography,
https://doi.org/10.1007/978-3-030-61895-7_2

and in the sand beaches of uninhabited islands far away in the ocean. By now, garbage rivers, islands and even continents have appeared on the surface of the ocean—where ocean currents converge. In the North Pacific Ocean, the area of the garbage continent is three times that of Spain with Portugal combined, and the mass of floating plastic is 60 times that of zooplankton. Much worse than that, only 0.5–1% of all plastic floats on the surface!

For a long time, it was believed that plastic waste had mainly an aesthetic effect. However, a number of studies have shown that plastic garbage has a significant impact on shipping, primarily in terms of safety of navigation. Biogeochemical processes occurring with plastic in sea areas have even more significant negative impact. Hydrolysis, photolysis and microbiological redox reactions destroy the polymer base of plastic, and it is more actively exposed to weathering and deformation. This results in the formation of fragments of various sizes, including microscopic ones. The process of plastic destruction in marine areas takes from several months to the first years. Thus, an ever-increasing number of fragmented polymers, called microplastics, should be noted [1]. Different researchers define the concept of "microplastics" in different ways. For example, Gregory and Andrady [4] believe that microplastics are hardly visible particles that freely pass through a mesh filter with a mesh diameter of 500 μm, but are trapped by a mesh filter with a mesh diameter of 67 μm, while larger particles are called mesomusors. Dimensional gradation of microparticles in terms of less than 5 mm is also common [5].

However, to date, the methodology and terminology of the experiments and data collected are far from uniform. The division into micro and macro plastic contamination can be considered relatively established [6]. Macroplastics is usually understood as "more usual" plastic garbage", which can be observed practically in any part of the World Ocean land due to almost universal human settlement, cheapness of plastic production, its resistance to degradation, low coefficient of waste utilization, as well as geophysical reasons: wind and currents (Fig. 2.1). The term microplastics has been used relatively recently. It is commonly understood as artificial polymer particles with dimensions less than 5 mm. The lower boundary is

Fig. 2.1 Differences in microplastic size

often determined by the purpose of the study or by the equipment used. In some works, particles smaller than 0.5 mm are also called nanoplastics.

Trends and threats of plastic pollution in the anthroposphere are recognized and described in basic documents UNEP, NOAA, Green Peace, etc. Certain measures are also taken at the level of national programmes.

In the case of microplastics, estimating the degree of contamination, the impact on biota and, in particular, on human beings is complicated by the labor-intensive and relatively high cost of conducting research. In addition, the global nature (Fig. 2.2) of this pollution requires harmonization of approaches, measurement units and identification methods.

2.2 Main Processes of Formation and Degradation of Microplastics in the Marine Environment

There are three main categories in which microplastics (MP) in the world ocean is being studied [6]:

- **biological and biochemical**: the presence of MP in the digestive and excretion systems of zooplankton, fish and animals, as well as laboratory studies of the effect of MP on the functioning of living organisms, study of the processes of sorption-desorption of heavy metals of PCBs and other hazardous pollutants;
- economic-social: study of production, transport and processing volumes plastics, recycling methods, etc.;

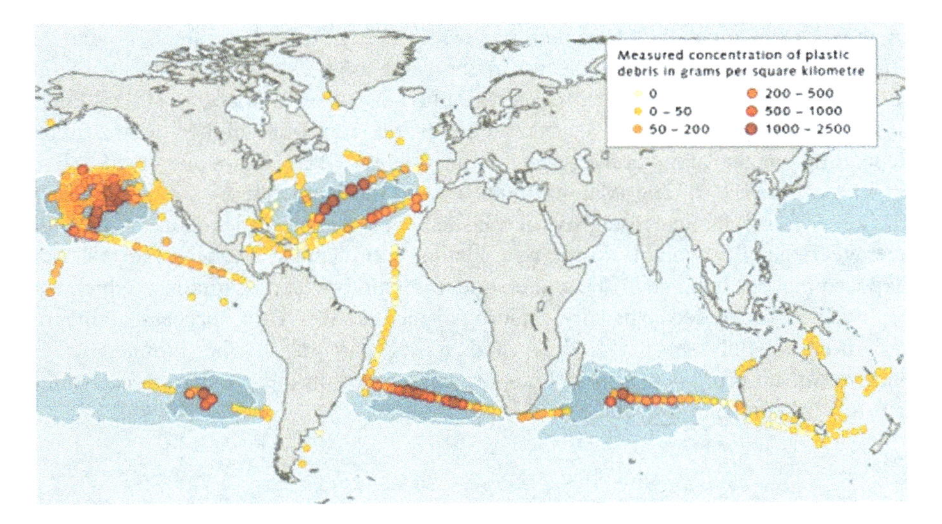

Fig. 2.2 Plastic waste concentrations in the world's oceans, in grams per km^2 [19]

- physical: processes and mechanisms of plastic destruction, its transport in rivers and seas;

  and in the ocean, shore-based exchange processes, numerical modelling.

In the physical part, the main idea is to understand the principles of particle movement at the micro level, i.e. transfer with currents and winds, deposition in the water column, excitation and discharge to the shore, accumulation depending on environmental conditions, the influence of shape and density and, finally, the processes of generation of microplastics from macroplastics [7]. The main difficulty in the physical description of MP particle transport is, firstly, the variety of properties of the particles and, secondly, the variability of these properties over time of plastic's "life" in the environment. Indeed, the initial plastics densities already vary in a wide range—from 0.01–0.05 kg/m^3 (in foam forms) to almost 3 kg/m^3, the sizes of MF particles detected in the marine environment are characterized by distributions, and their forms are extremely diverse [7]. In this case, the integral density of particles changes over time as a result of aging material, biofouling and/or aggregation with organic or sediment particles, particle size decreases due to fragmentation caused by mechanical loads and UV-radiation. Describing the behaviour of objects of this kind is a difficult task. especially in the natural conditions of the coastal sea zone.

Two main processes leading to the formation of microplastics should be highlighted: direct ingestion into the marine environment (some fragments (micro- and nanoparticles) used in consumer goods enter the waste water area, e.g. pellets in cosmetic scrubs or industrial synthetic abrasives) and weathering of larger debris in the marine and coastal environment. According to research [4], the main mechanism of microplastic generation is the destruction of larger plastic materials entering water from land. Plastic waste is present everywhere in beach areas, surface waters and deep water, but the rate of weathering in each of these areas varies significantly. In the coastal zone, the dominant process is the temperature effect. Given the relatively low specific heat capacity of the sand (664 J/kg K), the surface of the sandy beach and the plastic debris on it can be heated in summer to +40 °C. Photo oxidation decomposition is accelerated several times at higher temperatures depending on the activation energy of the process (AE). For example, if AE is 50 kJ/mol, the rate of degradation doubles when the temperature increases by only 10 °C. The mechanical integrity of plastic is invariably dependent on its high average molecular weight, so the degradation significantly weakens the material. Exposed plastics become brittle and disintegrate into powdery fragments, which can be further degraded (usually microbial-mediated). This process converts carbon-based polymers into CO_2 (and forms part of marine biomass). The decomposition process is completed—organic carbon in the polymer is converted and full mineralization occurs.

Polymer degradation agents

Degradation of polymers can be classified according to the agents causing it [1]:

- biodegradation associated with the activity of living organisms, primarily germs;
- photodegradation caused by solar activity;
- Thermo-oxidation that occurs under the influence of temperatures;
- hydrolysis—reaction with the aquatic environment.

All these processes lead to significant impacts on biota and the marine environment in general. Persistent organic pollutants (POPs), which are everywhere present in seawater at very low concentrations, are absorbed by microplastics through interchange. It is the hydrophobicity of POPs that enhances their concentration in microplastics, reaching values several orders of magnitude higher than the natural background. Also high risk of contamination is connected with bioavailability of POPs concentrated in polymers, which get into food chains by absorption of plastic fragments by marine biota. Microparticles and nanoparticles reach the size of phytoplankton, which is part of the diet of some representatives of zooplankton, such as Pacific krill. Studies [8] have noted that Pacific krill (Euphasea Pacifica) absorbs feed algae along with polyethylene granules of the appropriate size, with no apparent preference for food. However, no such studies have been carried out with plastic fragments containing POPs; it is also not known whether there are chemotactic or other warning signals that prevent them from being ingested (as opposed to POP-free granules) by at least some of the species at risk. Information on the bioavailability of POPs sorbed in different species after absorption of contaminated microplastics is extremely rare. The bioavailability of polycyclic aromatic hydrocarbons (PAHs) from technogenic particles such as tire fragments and diesel soot that have been placed in intestinal fluid has been identified in sea worms feeding on bottom sediments [9]. Surfactants contained in the intestinal fluid of bottom detritophages may increase the bioavailability of POPs in these species. On plankton, which is low in body weight, large quantities of POPs have toxicological effects. In this case, the dose depends not only on the volume of the micro-particle, but also on the time of its presence in the body and the kinetics of the transition of POPs from it to the tissues of the zooplankton. In large marine species such as the great petrels (Puffinus gravis), the amounts of contaminated plastic particles entering the body and POPs (polychlorinated biphenyls (PCBs), DDE, DDT and dieldrin) in adult fat tissue were positively correlated. Data on microplastic transfer coefficients of POPs at all marine trophic levels are unknown.

As for plastics with a high molecular weight, they are not subject to noticeable biodegradation, since the species of microorganisms that can metabolize polymers are rare in nature. This is particularly true for the marine environment, except for biopolymers such as cellulose and chitin. However, the work [10] identified several strains of microbes capable of decomposing polyethylene (Rhodococcus ruber for strain C208, Brevibacillus borstelensis for strain 707) and PVC (Pseudomonas putida). In a concentrated liquid culture, the actinomycetes of Rhodococcus ruber (strain C208) were processed up to 8% polyolefin in a laboratory into a dry mass during 30 days of incubation. The lacquers secreted by this species reduced the average molecular weight of the polymer. However, this process is almost impossible in soil and marine environments, as these microorganisms are not found in high concentrations and, moreover, there are always sources of easily assimilable nutrients in nature.

According to the data [1] it is necessary to note toxic impact of plastics, which can be attributed to the following factors:

Residues of monomers present in the plastic or toxic additives used in its production may be leached as a result of absorption of the plastic by marine animals. The potential toxicity of phthalate plasticizers used in PVC production has been widely discussed in the literature [11].

- Toxicity of some intermediate products of partial degradation of plastics. For example, when burning polystyrene, styrene and other aromatic compounds may form, and partially burnt plastics may contain significant levels of styrene and other aromatic compounds.
- POPs present in seawater are gradually absorbed and concentrated in plastic fragments. Thus, on the one hand, the plastic waste contributes to cleaning the marine environment from the pollutants dissolved in it. On the other hand, when it enters the body, these fragments become bioavailable and threaten the life of marine organisms.

The risk associated with high concentrations of POPs is of particular importance. Seawater generally contains some amount of chemicals such as pesticides and industrial chemicals that enter the ocean with wastewater [12]. POPs [polychlorinated biphenyls (PCBs), polybrominated biphenyl esters (PBBEs) and perfluorooctane acid (PFC)] have a significant polymer-water distribution coefficient in favour of the polymer. This factor can be approximated by the lipid and water distribution factor. However, this may lead to serious underestimation of the polymer-water distribution coefficient for some POPs [13]. It was important to discover that the desorption of the pollutant (back into water) was very slow and that even the sludge desorbed the phenanthrene faster than polymer fragments. Higher values for polymers were also pointed out. For example, in [14], this factor was 27,000 l/kg for polyethylene. The high variability of the experimental values of the distribution coefficient can be explained by differences in water temperature, the degree of crystallinity of the polymer and non-equilibrium effects. Some studies confirm that plastic can also store metals [15]. These results were unexpected because plastics are hydrophobic, but functional groups can form on an oxidised surface that can bind metals.

Thus, plastic waste, including microplastics, represents a significant threat to the marine environment. This is without regard to the effect of the physical accumulation of plastic particles in the bodies of marine facilities, which leads to digestive disorders and subsequent death. Clearly, the problem of microplastics requires further and detailed study. This is especially true for the coastal and marine waters of the Russian Federation. Research here so far has been sporadic; however, the first attempts at systematic observation of microplastics in coastal waters have been made in the waters of the Amur and Ussurian Gulfs (Sea of Japan) [16, 17].

Even the most modern numerical models are still unable to predict the transfer of microplastics by ocean currents. The reason for this is the constant change in the physical properties of particles over time in the "life" of the environment. Destruction into smaller fragments changes their size, biofouling changes the integral density of particles, and their shapes are infinitely diverse: from tiny fibres and thin threads, elastic films and flat flakes to three-dimensional fragments.

The tasks of numerical modeling currently address the global problems of plastic pollution transport at the level of the oceans and seas, but in order to develop more reliable predictive models it is necessary to accumulate data on the shore, as well as to parameterize MP sources and exchange processes with the shore [18]. Most field studies provide data on instantaneous MP concentrations in water, regardless of the current hydrometeorological conditions, prevailing currents and without indicating the possible temporal variability of MP concentrations. Another feature of plastics in general and microplastics in particular is their exceptional longevity in natural conditions. In fact, this is what mankind has created and produces plastics for: chemically inert, wear-resistant, not afraid of water and natural temperature changes. Duration of "life" of plastic bottles in the ocean is estimated at 450 years, and children's diapers and disposable diapers—at 500–600 years. Thus, all plastic produced since the beginning of the "plastic era" in the mid-1950s is still with us, it has not had time to decompose. Just imagine: if you accept that generations of people are replaced after 25 years, the diaper thrown into the sea will "live" 20 generations, that is, noticeably longer than its first owner! And then? And then it will turn into millions of microplastic particles. No matter how small the pieces of microplastics are, they retain their inert, long synthetic polymolecules, remaining inedible for the decomposing other organic residues of bacteria and microorganisms. But fish, birds, mollusks, sponges—on the contrary, they prefer microplastics to natural food. For example, albatrosses on islands in the North Atlantic willingly feed their chicks with these bright colored pieces, taking them for food… Many species of zooplankton, and behind them, fish absorb microplastics, thereby allowing them to penetrate the food chains and climb up to our table. As plastics age, numerous additives, dyes, stabilizers, etc., are released into the environment (and into the organisms into which they have been ingested) and used in the manufacture of plastics. In addition, the roughened surface of microplastics particles is also capable of collecting toxins from the environment. All this also has a negative impact on living organisms. Further, the surface of microplastics particles has proved to be a very convenient new habitat for microorganisms. Their colonies are rapidly repopulating the available surfaces and… and they can easily travel vast distances on these ships. Will this path be accessible to potentially pathogenic species as well? Obviously, it's impossible to cleanse the environment of an unimaginable amount of microplastic particles. It is clear, that modern civilization is not able to refuse use of synthetic polymers. However it is quite possible to reduce to a minimum use of "disposable" plastic—disposable ware, polyethylene bags, packages, etc. which make almost half of weight (!) of plastic made in the world. Another very real opportunity to reduce the severity of the problem—sorting and further processing of plastic in other products: this way turns the "eternity" of plastic again in its undeniable advantage.

2.3 Conclusions

A review of published information on the causes, sources and current levels of microplastics contamination of the world's oceans is given.

It has been shown that micro- and macro plastics, as particularly persistent and long-lasting oceanic pollutants, have now become a global problem, both trans-boundary and social, economic and environmental.

The main processes of plastics destruction in sea areas are described, and the main categories in which microplastics behaviour in the world's oceans is investigated are given.

References

1. Kozlovskiy NV, Blinovskaya YY (2015) Microplastics—macroproblems of the World Ocean. Int J Appl Basic Res Part 1:159–162
2. Hinojosa I, Thiel M (2009) Floating marine debris in fjords, gulfs and channels of southern. Chile Mar Pollut Bull 58:341–350
3. UNEP (2016) Marine plastic debris and microplastics—Global lessons and research to inspire action and guide policy change. United Nations Environment Programme, Nairobi, 252 p
4. Gregory MR, Andrady AL (2003) Plastics in the marine environment. Plastics and the environment. Wiley. ISBN: 0-471-09520-6
5. Barnes DKA, Galgani F, Thompson RC, Barlaz M (2009) Accumulation and fragmentation of plastic debris in global environments. Philos Trans R Soc:364
6. Bagaev AV, Chubarenko IP (2019) Integrated research of microplastics in the World Ocean. In: Conference "KIMO-2019" section "Plenary reports of young scientists"
7. Chubarenko I, Bagaev A, Zobkov M, Esiukova E (2016) On some physical and dynamical properties of microplastic particlesin marine environment. Mar Pollut Bull 108(1–2):105–112. https://doi.org/10.1016/j.marpolbul.2016.04.048
8. Berk SG, Parks LS, Tong RS (1991) Photoadaptation alters the ingestion rate of paramecium bursaria, a mixotrophic ciliate. Appl Environ Microbiol 57(8):2312–2316
9. Voparil IM, Burgess RM, Mayer LM, Tien R, Cantwell MG, Ryba SA (2004) Digestive bioavailability to a deposit feeder (Arenicola marina) of polycyclic aromatic hydrocarbons associated with anthropogenic particles. Environ Toxicol Chem 23:2618–2626
10. Sivan A (2011) New perspectives in plastics biodegradation current opinion in biotechnology. Curr Opin in Biotechnol 22(3):422–426
11. Latini G, De Felice C, Verrotti A (2004) Plasticizers, infant nutrition and reproductive health. Reprod Toxicol № 19(1):27–33
12. Wurl O, Obbard JP (2004) A review of pollutants in the sea-surface microlayer (SML): a unique habitat for marine organisms. Mar Pollut Bull № 48(11–12):1016–1030

13. Friedman CL, Burgess RM, Perron MM, Cantwell MG, Ho KT, Lohmann K (2009) Comparing polychaete and polyethylene uptake to assess sediment resuspension effects on PCB bioavailability. Environ Sci Technol №43(8):2865–2870
14. Lohmann R, MacFarlane JK, Gschwend PM (2005) On the importance of black carbon to sorption of PAHs, PCBs and PCDDs in Boston and New York harborsediments. Environ Sci Technol 39:141–148
15. Ashton K, Holmes L, Turner A (2010) Association of metals with plastic production pellets in the marine environment. Mar Pollut Bull 60(11):2050–2055
16. Blinovskaya Y (2005) Primorsky kray shoreline pollution monitoring methods and results. In: First international workshop on marine litter in the Northwest Pacific Region 14–15 Nov 2005. Ministry of the environment of Japan, pp 98–104
17. Blinovskaya YY, Vysotskaya MV (2012) Analysis of the marine garbage management system in the NOWPAP region (in Russian). In: Marine State University Bulletin. Series: Theory and practice of sea protection. 55/2012. Marine State University. Un., Vladivostok, pp 3–11
18. Chubarenko I, Esiukova E, Bagaev A, Isachenko I, Demchenko N, Zobkov M, Efimova I, Bagaeva M, Khatmullina L (2018) Behavior of microplastic sin coastal zones. In: Microplastic contamination in aquatic environments (Chapter 6). Elsevier, pp 175–223
19. http://worldoceanreview.com/en/files/2015/11/wor4_en_k2b_abb_2-19-500x284.jpg

Chapter 3
Current Problems of the World Ocean Pollution by Oil and Oil Products

3.1 Introduction

Oil is a complex mixture of many components. These components include linear, branched, cyclic, monocyclic aromatic and polycyclic aromatic hydrocarbons. In high concentrations, hydrocarbon molecules are highly toxic to many organisms. Oil also contains trace amounts of sulphur and nitrogen compounds, which are dangerous in themselves and can react with the environment, resulting in secondary toxic chemicals. The predominance of oil products in the world economy creates conditions for the distribution of large quantities of these toxins in populated areas and ecosystems around the world [1].

Crude oil production, especially offshore production, is steadily increasing by sea transport, as is the amount of oil products entering the sea. The total amount of oil and petroleum products entering the ocean annually is estimated by various researchers at 6–12 million tons.

Crude oil and its derivatives are an extremely complex mixture of many chemical compounds. It contains about 300 hydrocarbons, individual concentrations of which vary in volume from 10 to 4% to units of percent. The physical and chemical properties of oil vary greatly from field to field, and its different grades differ in the content of paraffin, tar and sulfur. Crude oils vary greatly in density, from very light (0.65–0.70 g/cm^3) to very heavy (0.98–1.05 g/cm^3). The viscosity of crude oil depends on its chemical and fractional composition, especially its resinosity, and also varies widely.

At annual volume of pollution of the World Ocean in 6–11 million tons, the sources of pollution are distributed as follows: sea transport (washing water, docking, leakages, loading and unloading operations, etc.)—35%; industrial runoff—13%; marine oil production—1.5%; river runoff—32%; atmospheric input—10%; natural sources of oil input—about 10%. A relatively small amount of oil entering the ocean directly as a result of its marine production and from natural sources attracts attention [2]. Analysis of space images of the sea surface shows that

K. Pokazeev et al., *Pollution in the Black Sea*, Springer Oceanography,
https://doi.org/10.1007/978-3-030-61895-7_3

areas of global oil pollution coincide with marine transportation routes and estuaries of the largest rivers. Large-scale pollution zones include not only shelf, but also some areas of the open sea part. Average pollution of such seas as Barents Sea and Baltic Sea. The Black Sea, Caspian Sea and Mediterranean Sea are several times more polluted than MAC (maximum permissible concentration).

3.2 Natural and Anthropogenic Sources of Oil in the Sea

Oil, as a natural product of biota decomposition, can participate in biogeochemical cycles of substance migration over millions of years. The volume of oil entering the ocean from the subsoil is about 0.6 million tons per year, with hydrocarbons being the bulk of it. In the World Ocean, 12 million tons of hydrocarbons are produced annually by photosynthesis. In the process of long-term evolution, mechanisms of assimilation and self-purification of water bodies from hydrocarbons have developed. However, in recent years there has been a significant excess of natural oil flow into the ocean. The assimilation capacity of a number of areas of the World Ocean has been exceeded. Another important point is that the quality of oil pollution has changed. As a result of washing operations on tankers not oil and oil products, but products of their processing—paraffin, asphalt and resinous components, oil residues—get into water bodies as a result of ship bilge water, from oil refining enterprises. Properties and composition of these contaminants differ from properties and composition of oils. High-molecular components of oil, products of polymerization and polycondensation of hydrocarbons, products of metal corrosion, etc. are concentrated in oil residues. The bulk of oil pollution enters the ocean in the form of oil-containing waters. According to some estimates, up to 75% of oil goes to the ocean in emulsified state. Crude oil deposited on the surface of the ocean as a result of accidental release is subject to mechanical and thermal effects, so the type of oil in the sea changes greatly over time. Crude oil in water can be in the form of films of varying thicknesses, emulsions, dissolved form and clots. Theoretically, oil can spread to monomolecular layers, but under real conditions oil films contain thousands of molecular layers. The size of oil particles in the emulsion is less than 3×10^{-2} mm. The solubility of oil depends on many factors and lies between 2 and 100 mg/l.

3.3 Oil Degradation Processes

Behavior of oil caught in the sea depends on its physical and chemical properties as well as on environmental conditions: wind and wave conditions, currents, amount of oil-oxidizing bacteria, air and water temperature.

When water gets to the surface, oil immediately begins to undergo changes due to the influence of various physical and chemical factors from outside. Such interaction of oil with the environment can be divided into several stages: spreading, evaporation, dissolution, emulsification, dispersion, oxidation and dispersion (Fig. 3.1). The fractions of oil are called components that can be separated from it during distillation. Different processes take different time and involve different fractions of oil—some occur with lighter components, while others with heavier ones, and under their influence the composition of oil changes. Thus, on the first day after a spill, the main processes are evaporation, dispersion, emulsification and spreading. After a week, the lighter fractions are volatilized and the remaining hydrocarbons are biodegraded. Oil also has the property of sorbing the solids in the water, forming aggregates and settling on the bottom as solid sediment. These processes can last from several weeks to several years.

Oil spills are caused by wind, surface currents and surface tension and gravity forces, forming areas covered with thin oil films thousands of molecular layers thick. In the absence of wind and currents 1 m^3 of crude oil can spread over an area of 50 m radius in 1.5 h, the thickness of the layer will be about 0.1 μm. Wind and surface currents are the factors that influence the dynamics of oil spreading. Thus, the greater is the wind speed, the faster is the slick spreading. It stretches along wind direction and thins quickly, after which it can break up into smaller slicks.

Features of oil and oil products spill modeling are considered in Matsenko [3].

Oil spillage is the result of gravity forces as well as viscosity and surface tension forces. The speed of this process depends on the ratio of surface tension coefficients at the water to air interface, oil to water and oil to air. Let the surface tension coefficients at the boundary of two media be indicated σ_{wa}, σ_{oa}, σ_{ow} (Fig. 3.2).

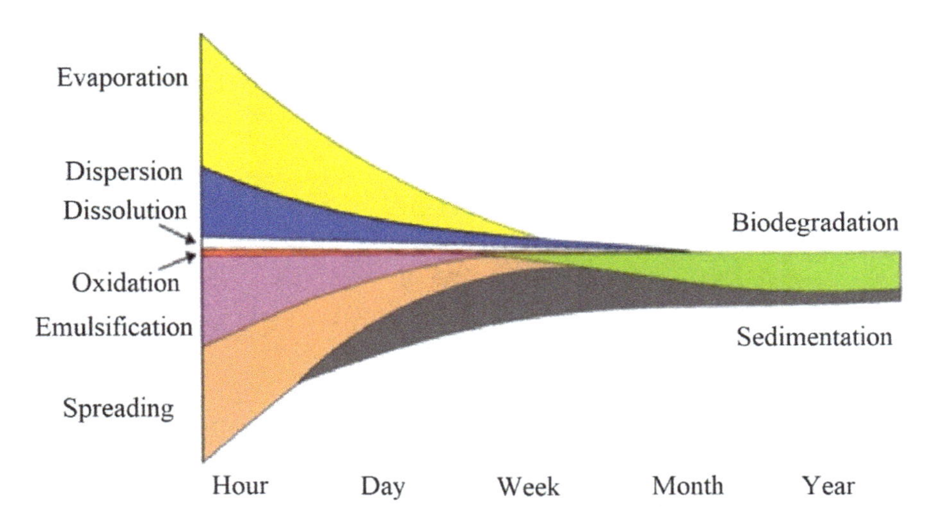

Fig. 3.1 Evolution of oil on water surface. http://www.medess4ms.eu/marine-pollution/attachment/biodegradation

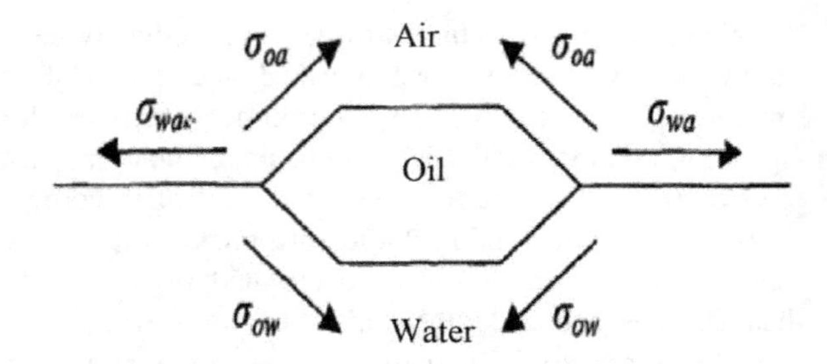

Fig. 3.2 Surface tension coefficient ratio diagram

By the sign of the spreading coefficient (or effective surface tension coefficient), calculated by the formula

$$\sigma = \sigma_{wa} - \sigma_{oa} - \sigma_{ow} \tag{3.1}$$

The following two modes can be distinguished at the media interface:

$\sigma < 0$—case study by Langmur [4]. An established lenticular oil slick is formed on the water surface with an average thickness of h, equal to

$$h = 2\,\sigma\,g\,\rho_o\,\rho_w\,\Delta\rho \tag{3.2}$$

where

ρ_w water density;
g free fall acceleration;
ρ_o oil density;
$\Delta\rho$ water and oil density difference.

Since the typical value of $\sigma = 2.5 \times 10^{-2}$ N/m, this case is uncharacteristic for the initial stage of oil spillage and can be realized over time by changing the values of coefficients σ_{oa}, σ_{ow}, due to changes in the composition of crude oil, which is associated with water and air and is subject to physical and chemical effects.

The very concept of the surface tension coefficient is directly related to isobaric potential, or Gibbs energy (G), by the formula:

$$\sigma = \left(\frac{\partial G}{\partial S}\right)_{p,T,n_i} \tag{3.3}$$

where

S phase contact area,
P pressure,
T temperature,
n_i phase concentrations.

Let's consider the "oil on water" system at constant temperature and pressure. Then:

$$dG = \sum \sigma_i \, dS \tag{3.4}$$

If the oil drop area increases by dS, the water-oil/oil/air interface will increase by the same amount and the water-air interface will decrease. It follows from here that:

$$dG = \sigma_{wo} dS + \sigma_{oa} dS - \sigma_{wa} dS \tag{3.5}$$

Then, for the spontaneous process of increasing the water-oil contact area (spreading) such that $dS > 0$, $dG < 0$, you can enter the Garkins spreading criterion:

$$K = \sigma_{wo} + \sigma_{oa} - \sigma_{wa} \tag{3.6}$$

Expansion possible at $K > 0$.

There are several models of oil spreading over water surface based on hydrodynamic equations. In the model proposed by Fay in 1969, the spreading process can be divided into the following phases [5, 6].

- inertia—occurs due to the balance of horizontal pressure gradient and inertia forces, the spot radius growth rate is determined by the ratio: $R = (\Delta g v t^2)^{1/4}$
- gravitationally viscous occurs due to the balance of horizontal pressure and viscosity forces, $R = (\Delta g V^2 t^{3/2} \nu^{-1/2})^{1/6}$
- surface tension phase—the main operating forces are viscosity and surface tension, $R = (\sigma^2 t^3 \rho^{-2} \nu)^{1/4}$
- the stain growth stop phase,
  where R—slick radius, V—spilled oil volume, t—time, Δ—relative water density determined by the formula $\Delta = \frac{\rho_w - \rho_o}{\rho_w}$, ν—kinematic viscosity of water, the expression for σ has been defined above, g—free fall acceleration, ρ_w and ρ_o—water and oil densities respectively.

The duration of the phases is calculated from Hoult's diagrams depending on the spill volume. However, the spot radius calculated using Fay's formulas is much smaller than that obtained experimentally. As was shown in Fay [6], the area of an oil spill depends on the time since the beginning of the spreading.

Even after oil has disappeared from the water surface, its harmful effects on ecosystems continue for a long time. In 1969, a coaster barge with oil ran aground near the Woods Hole Oceanographic Institution [7]. Scientists had a unique opportunity for scientific analysis of the consequences of oil impact on different types of coastal areas. Although the release of oil was brief, oil contamination on the marshes (marshy flooded areas) remained for 10 years. The area covered with large pebbles retained oil pollution for 2 years. The rocky area contained no pollution after 2 months.

3.4 Dynamics of Oil Slick on the Sea Surface

Let's take a quick look at the dynamics of the oil slick. If the spilled oil on the sea surface is not exposed to waves, wind and currents, then the main role in the formation of the oil spill area is first played by gravity. Then surface tension, viscosity and inertia forces begin to prevail. Depending on the ratio of forces that play a major role in the spread of oil, there are three stages in the process of oil spill formation. In the first stage (gravitational or inertial regimes of spreading), the forces of gravity and inertia play a decisive role. At the second stage—called gravity and viscosity regime—the determining role is played by forces of gravity and viscosity, at the third stage—by forces of surface tension and viscosity.

The duration of flow regimes depends on the initial volume of oil. The reason for stopping oil spreading is the decrease and change in the sign of the surface tension coefficient of the film on water. However, determination of this boundary is a very difficult task due to the fact that oil contains many organic substances that affect the value of the surface tension coefficient.

The displacement (drift) of the oil film as a whole as opposed to spreading is determined by external forces: wind, currents, surface waves. If the drift of the oil film occurs under the influence of wind and currents not caused by direct action of wind, the drift velocity is equal to the arithmetic sum of velocities.

$$U = U_W + 0.56U_C \tag{3.7}$$

where U_W—is the speed of wind drift, U_C—is the speed of drift caused by currents. Wind drift plays a decisive role in film drift. The wind drift velocity is about 3% of the wind speed. Due to the Coriolis force, the drift direction deviates from the direction of wind speed, the angle of deviation is between 0 and 100°.

Once the spillage has stopped, further growth of oil contamination will be caused by turbulent diffusion. Diffusion of oil contamination differs from diffusion of conservative passive impurities. Non-conservativity of oil film is caused by evaporation, oil dissolution and biodegradation; in the process of diffusion the chemical composition of the film changes. That is why it is necessary to approach with caution to transfer the results of experiments on horizontal turbulent diffusion with different dyes to the oil slick diffusion. It should be noted that in case of oil film there is horizontal diffusion, and for oil in dispersion state—diffusion will be three-dimensional.

Surface waves have a significant impact on oil pollution. Waves are the cause of wave drift, cause mixing of oil in the collapse of the waves, change the thickness of the oil slick over different parts of the wave. Wave motion creates a wave current in the surface layer due to non-linear effects. The velocity of the flow is determined by the Stokes formula.

The study of oil drift under the influence of waves and the combined effect of waves and wind has so far been conducted only in laboratory work. It turned out that in this case the wave drift velocity exceeds the value of Stokes velocity. Wave drift velocity in all cases is less than 20% of wind drift velocity. It should be noted that these results were obtained in laboratory conditions at low accelerations. The ratio between pure wind drift velocity and wave velocity depends not only on the wind speed but also on the stage of wave development.

There are a number of mathematical models that describe the physical and chemical kinetics of oil at sea:

1. The model of physical and chemical transformations of oil taking into account its evaporation, emulsification, sedimentation and microbiological phenomena [8, 9].
2. Oil Spill Spread Model [10].
3. Upper sea layer shear diffusion model [11].
4. Model of the drift of an oil slick under the influence of the wind force of a drive layer of the sea [12].

The surface of the ocean under certain wind conditions is covered by a significant number of crashing waves, which affect the mixing of oil and the dispersion of oil slicks. In turn, oil slicks reduce the frequency of wave crashes. Wave collapse on the oil slick leads to mixing oil in the water column, the buoyancy of the oil particles causes its particles to rise to the surface. The balance of forces that lead to the deepening and lifting of oil, determines the time of its particles in the water. The direct influence of pollutants, particularly oil, on the collapse of waves is maintained until the integrity of the surface films is broken. In strong winds, the surface slick is destroyed and an emulsion of oil in the water is formed in the layer of wind mixing, whose effect on the collapse of the waves is negligible. The increase in viscosity in the near-surface layer due to the formation of the emulsion leads to a slight weakening of the high-frequency components of the wave spectrum. Oil penetration into depth is accompanied by spatial differentiation of oil particles by size and oil hydrocarbons by chemical composition [13].

Waves can change the thickness of the oil slick above them. Measurements of the oil layer thickness above the wave surface (if the layer thickness is more than 1 mm) in laboratory and field conditions showed that the oil layer becomes thicker above the wave crest and decreases above the troughs. This effect was established in case of waves without wind. Studies show that friction stress and pressure in the wind flow can lead to a redistribution of the oil layer over the wave surface. A slick with large spatial dimensions leads to changes in the characteristics of the boundary driving layer—the roughness of the water surface, the coefficient of resistance, the friction speed, and the wind speed in the driving layer changes. As a result, energy inflow from wind to surface waves and currents changes, gas and moisture exchange of boundary layers is disturbed. The use of surfactants in agriculture to reduce evaporation from small water bodies is a well-known fact. Thus, film contamination of large sea areas can significantly disrupt the mechanisms of interaction between the atmosphere and the ocean.

3.5 Conclusions

The information on peculiarities of the world ocean pollution by oil and oil products is given, the dynamics of their distribution depending on their composition and sources is shown.

The nature of oil and oil products degradation processes in sea water is analyzed.

The dynamics of oil slick on the sea surface under the influence of wind and waves is considered and its influence on the mechanisms of ocean-atmosphere interaction is estimated.

References

1. Kartamysheva ES, Ivanchenko DS (2018) World ocean pollution by oil and oil products (in Russian). Young Scientist 25:20–23
2. Pikovsky YuI (1993) Natural and technogenic hydrocarbon flows in the environment. Moscow State University Publishing House, Moscow, p 202
3. Matsenko SV (2009) Modeling of oil and oil products spills. In: Matsenko SV, Dunets LG (eds) State regulation in the field of prevention and response to spills of oil and petroleum products at sea and inland water areas: the materials of the scientific workshop 21–23 April 2009. Novorossiysk, FGOU VPO "Admiral F.F.Ushakov MGA", pp 34–36
4. Langmur I (1933) Oil lens on water and the nature of the monomolecular expanded films. J Chem Phys 1:756–776
5. Fay JA (1971) Physical processes in the spread of oil on a water surface. In: Proceedings of the joint conference on prevention and control of oil spills. American Petroleum Institute, Washington, DC, pp 463–467
6. Fay JA (1969) The spread of oil slicks on a calm sea. In: Oil on the sea. Plenum Press, New-York, pp 53–63
7. Revel P, Revel C (1995) The environment of our habitat. In: Water and air pollution. World, Moscow, 296 p
8. Murray SP (1979) The effects of weather systems, currents and coastal processes on major oil spills at sea. In: Pollution transfer and transport in the sea, part II, pp 169–225
9. Leech MV (1991) Development of an oil and chemical spill simulation model for the NW European Continental shelf and the Western Mediterranian/Adriatic Seas. Warren Spring Laboratory, Stevanagem Report LR 810
10. Milgram J (1978) The role of physical studies before, during and after oil spills in the wake of the Argo Merchant. In: Symposium, Center for Ocean Management Studies University of Rhode Island, pp 5–85
11. Murty TS, Khandekar ML (1993) Simulation of movement of oil slicks in the Strait of Georgia using simple atmospheric and ocean dynamics. In: Proceedings of the 1973. conference on the prevention and control of oil spills, pp 42–58
12. Eremeev VN, Ivanov LM (1987) Tracers in ocean: transfer parameterization, numerical modeling of dynamics. Kiev, Naukova Dumka, 145 p
13. Trukhin VI, Pokazeev KV, Kunitsyn VE (2005) General and ecological geophysics - M. Fizmatlit, 576 p

Chapter 4
Numerical Modeling of the Hydrocarbon Spot Shape on the Water Surface

4.1 Introduction

The accuracy of spill characterization estimates depends on whether adequate information about the source of the oil discharge and the hydrometeorological conditions in the spill area is available to the predictor. The choice of model is determined by the task to be performed and each model needs to be customized or adapted to the specific conditions of the region in which the model will be used.

The success in predicting the spread of oil at sea is determined not only by the ratios used to calculate the characteristics of a spill, but also by the ability to use the results of meteorological forecasts, marine dynamics forecasts, including calculations of sea wind waves and ice cover characteristics supplied by models adapted and verified for the region under consideration.

In order to plan actions to prevent and respond to possible oil spills, it is necessary to be able to predict the consequences of these spills: possible routes of flow and accumulation of oil, the impact of oil on natural objects, etc. Information about the behavior and nature of a spill in one or another case allows for the quickest possible introduction of a spill control mechanism, thereby reducing the number of spilled oil products and reducing environmental damage. The system for predicting the consequences of oil spills is based on modern methods of mathematical modeling, taking into account the hydrodynamic and climatic characteristics of the spill zone.

The complexity of modeling oil spills is that it is necessary to take into account many factors: wind speed and direction, surface currents; physical and chemical properties of the fractions that make up the oil (boiling point, viscosity, molecular weight); evaporation and, consequently, changes in the density, viscosity and volume of the slick. It is also necessary to keep in mind the shoreline boundaries, bottom topography in case of shallow water and the presence/absence of ice cover on the surface.

K. Pokazeev et al., *Pollution in the Black Sea*, Springer Oceanography,
https://doi.org/10.1007/978-3-030-61895-7_4

Today all developed countries use mathematical-models of oil spills, implemented as a corresponding software product. The most famous of them are OILMAP and GNOME (USA), COZOIL and OSCAR (Norway), OSIS (UK), SPILLMOD and OilMARS (Russia), PADM (Sweden). In order to use all listed models, integration with hydrometeorological forecast output in the area of interest is required. It should be noted that all models are commercial products.

Despite the fact that there are many programs that allow calculating oil's behavior on water, none of them takes into account all external factors. The characteristics of the spill source also have a great influence on the slick behavior. Accordingly, models describing both separate and different spill sources have been developed. For example, Reed [1] describes the formation of oil slick at oil pipeline breakthrough and does not consider the behavior of the slick after its formation.

The SL Ross Oil Spill Fate and Behavior Model (SLROSM) allows the simulation of different emission sources. The starting region, slick thickness and oil properties are determined for different spill scenarios, and then the basic model of oil behavior is used to calculate the oil behavior on the water surface. SLROSM also allows the simultaneous simulation of the behaviour of several slicks, which are calculated separately, as different slicks have different wind dynamics due to different formation times.

Many modern models rely on early work by Fay [2] and Mackay [3] with some modification. For example, Belore R uses a modification that takes into account changes in oil density and the determination of oil tension at the point of spill. As for vertical dynamics, Delvigne [4].

For a more adequate description of the oil slick behavior, three-dimensional models describing the oil slick movement on the water surface and vertical interaction with the water column are considered. Early models Fay [2], Mackay [3] determined the temporal dynamics of an oil slick, not its shape (the shape of a circle or ellipse was supposed to be a circle). Johansen [5] and Elliott [6] suggested a hypothesis that the observed spreading of the slick occurs as a result of oil droplets sinking and floating. On the basis of what the kinetic model of vertical mixing of oil droplets under the action of waves was developed [7].

The main characteristics of oil, destroying waves and water column are gathered in a single "blending factor", which determines the share of oil distribution in the film and water column. It includes oil viscosity, dependence on the wave height, stress coefficient between the surfaces oil density. This approach summarizes the Delvigne surfacing model.

To determine the location and configuration of the slick often used the method of an ensemble of Lagrangian particles [8]. They simulate the movement of parts of a stain on the surface of a reservoir with account of wind, currents, wave and turbulent diffusion. The velocity of horizontal displacement of Lagrangian particles can be described by the following ratio:

$$U_a = U_{current} + U_{drift}(W) + U_{wave} + U_{dif}$$

W—wind speed over slick, $U_{current}$—velocity of the flow, calculated in the current model, $U_{drift}(W)$—drift velocity, U_{wave}—transfer velocity due to excitation, U_{dif}—random diffusion correction. This formula is not universal, but is geo-referenced and its models of ocean and atmosphere dynamics. Equations of Fay rare often used to calculate the oil slick area.

Here is a brief description of the most common oil spill models developed in different countries.

4.2 OILMAP Model (USA)

The OILMAP model predicts the spill of oil. Developed by Applied Science Associates (ASA) and used by such companies as BP, Royal Dutch Shell, ExxonMobil and others. It can be used in operational mode, allowing the input of wind and sea dynamics data. It operates on the basis of a geo-information system, also developed by ASA, but can also be used with other GIS. Can predict both the trajectories of continuous slicks and those already broken into smaller spots. Considers evaporation, emulsification, oil current capture, shoreline. The source of pollution in the program can be set in the form of a point, several points or some area.

Currently, the OilMARS model works in 3D, calculating the intra-water propagation of the dispersion plume and the oil submerged in water. The model is capable of calculating the appearance and spread of secondary oil contamination on the water surface and sea floor contamination. In addition, it provides for the possibility of taking into account the influence of cohesion and drifting of the ice cover on the spread of pollution, as well as oil penetration on the upper surface of the ice cover and under the ice as a result of compression of drifting ice.

A prolonged oil spill is represented as a large number of small discrete spills—portions or spillets—which are delivered to the water surface at a certain periodicity from a pollution source, depending on the spill speed, generally variable in time. Each spillet has a set of parameters: the coordinates, area, density and viscosity of oil, the amount of oil on the water surface, the amount of vaporized oil, etc. All parameters of spillets depend on the time of the location of the spillet on the water surface. The approach used in the model makes it possible to take into account the spatial heterogeneity of the oil slick, i.e. at each point of time on the water surface are spilletons with different density, viscosity, mass and area.

The Lagrangian approach is used to calculate the transfer of spillways and the calculation grid from the water circulation model is used. Spillette transfer occurs under the influence of wind, currents and waves. To calculate the processes of oil transformation Euler's approach and the grid with high spatial resolution, depending on the initial mass of spillets (Fig. 4.1B) is used. When the spillet

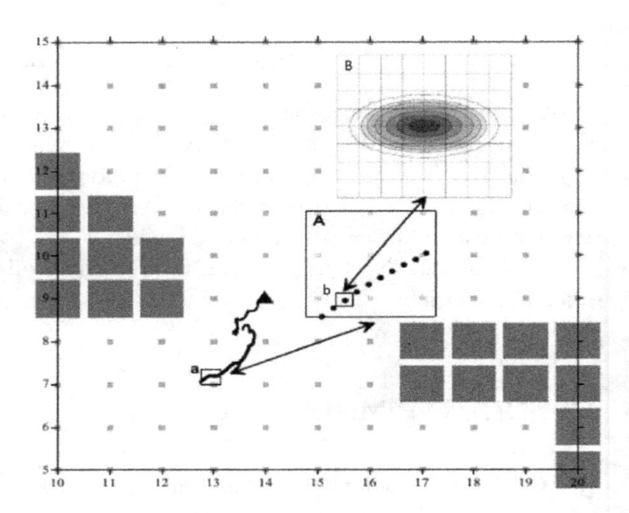

Fig. 4.1 Scheme of application of Lagrangian-Ailer approach to oil spill transfer and transformation description. The areas marked with lowercase letters correspond to the inserts marked with capital letters. On axes—grid coordinates

approach to the shore or brazing is used conditions of nonflowing and slip. In this case, a section of polluted coastline is fixed.

The process of evolution of each spillet goes through two stages: spreading and transformation. To calculate the spreading of oil to the state of the film, the model uses a standard approach [9]. Due to the small volume of the spill, the process of spreading takes only a few hours, during which the evaporation of light fractions of oil and a corresponding decrease in the volume of the spill. In addition, the cohesion of the ice sheet is taken into account. In this case, the spreading spillet is carried over the sea surface. After the spreading is completed, the slick is adapted to a rectangular calculation grid, taking into account the wind effect.

The OILMAP model is widely distributed and used for calculations in the water areas of the eastern coast of the USA, Australia, Brazil and others. Comparison of modeling results with satellite images showed that the model gives quite satisfactory results [10].

4.3 Model OilMARS (RF)

The OilMARS (Oil Spill Model for the Arctic Seas) was developed by the Arctic and Antarctic Research Institute (AARI). It calculates the transformation and movement of an oil spill, both instantaneous and long-term, from a fixed or moving source. In addition, this model can be used to calculate the behavior of oil in the water column and the distribution of dispersion particles. A prolonged oil spill is represented as discrete spills (spillways) coming to the water surface with a given period of time. Each of them is defined by a number of parameters, such as viscosity, coordinate, area, oil density, etc., independently of the others. The initial masses of spillets depend on the intensity of the source, which can be given a dependence on time.

The model allows to consider the presence of ice cover. If the ice cover is more than 5 points solid, the influence of wind waves is not taken into account. If the drift velocity of the ice is greater than the spillet's own velocity, it is assumed that they move at the same speed. When the spike approaches an area of high ice cohesion or the shoreline, the area of the spike decreases with the protection of the mass and the increase in film thickness.

The interaction of the spillet with the ice cover can be divided into two cases. If the ice is more than 9.5 points solid and its compression rate is less than 0.12 m/s, the spillet is under the ice. At higher speed part of the cutlet, determined by a random factor from 0 to 1, is splashed on the ice surface.

The wind component of the spot movement speed, as mentioned above, is assumed to be 3% of the wind speed and deviates from the direction of the spot by 15° clockwise. Wind wave transport is calculated from the wind wave model, and the transport speed is assumed to be 1.5% of the wind speed. When calculating the oil spill evolution, the Euler method and a high spatial resolution grid are used [11, 12].

The disadvantages of the model are the following: in weak winds, the neighboring spillways can overlap many times, which introduces an error in the calculation of the total area and spot thickness, and these are very important parameters when planning a spill.

The model is positioned as WOSM (world wide oil spill model), but it works effectively only in areas for which hydrometeorological information is available on-line.

4.4 Model Seatrack Web (Sweden)

The model has been developed to predict the spread of oil spills in the Baltic Sea, off the Swedish and Danish coasts and in the Skagerrak and Kattegat Straits. It is applicable for different types of hydrocarbons, from gasoline to asphalt, as well as for algae and frozen, eroded oil. Slick force values are calculated using HILRAM weather modeling and HIROMB ocean modeling. Seatrack Web is based on the use of Lagrangian particles: an oil slick is represented by a set of particles and the trajectory of each of them is calculated separately taking into account the external fields acting on it, changing in time. Each particle is described by a set of parameters that also change over time: its coordinate, density, mass, volume, etc., and the trajectory of each particle is calculated separately taking into account the external fields acting on it.

Modelling of the interaction with ice is based on the fact that the ice cover reduces the excitement on the water surface, thereby reducing the influence of the Stokes flow, as well as reducing the free water surface. This increases the thickness of the film. When the ice concentration reaches a critical value, it is believed that the speed of its drift determines the speed of slick movement.

The movement speed of the stain is added up to the following values:

- Own speed, defined by hydrodynamic model
- Speed caused by excitement (Stokes)
- Horizontal spreading caused by gravitational forces
- Emulsification
- Turbulent phenomena
- Surfacing and sinking of particles.

This process is also influenced by the very source of spreading. The model presents three types of sources: particles start moving from a single point, are evenly distributed along the line, or are evenly distributed within triangular or quadrangular regions. The time of spillage can also be different: either it occurred at a certain point in time, or the particles are emitted with a certain period. The source can be located both on a surface, and under water (Liungman O., Mattsson J., Massmann S. Scientific documentation Seatrack Web. Physical processes, algorithms and references [13].

The disadvantages of the program include the following properties: the ability to simulate the spreading of only 2 or 5 days forward or backward, the simulation is carried out only on the plane, not in the water column, the concentration of petroleum products is not indicated graphically.

4.5 Model SpillMod (Russia)

SpillMod [14, 15] takes into account the main processes occurring in the spill under the influence of the environment (spreading, transport and deformation under the influence of wind and currents, evaporation of the components of the spill, natural dispersion and changes in the properties of the spill over time). The model is based on the use of Navier-Stokes equations averaged in vertical coordinates and allows the necessary calculations, including for the initial stages of the process, in areas of complex geometry, with free and contact boundaries, the presence of ice cover.

The equations were obtained by a small parameter perturbation method from the three-dimensional problem for the flow of a thin layer of light incompressible Newtonian fluid on the surface of a more dense substrate. At the boundaries of oil-air and oil-water the conditions of stress continuity and kinematic conditions are set. Similar to Seatrack Web, SPILLMOD uses Lagrangian particles to describe the movement of oil contamination. The Eulerian grid is used to describe the external fields.

A distinctive feature of this package is the solution of hydrodynamics equations integrated on the vertical coordinate in areas with arbitrary contact boundary geometry, which allows you to calculate the configuration of contamination at any given time. The model also allows to take into account the use of dispersants and booms [16].

4.6 The GNOME Model (USA)

The program calculates the trajectory of the slick using the Euler and Lagrangian models: the hydrocarbon slick is presented as a set of Lagrangian particles moving in the Euler velocity field. The project is opensor-based, meaning everyone has access to the code. The disadvantages include the fact that the speed and area of slick propagation in this program does not depend on the spill volume, which does not correspond to the real nature of the liquid movement.

4.7 Conclusions

As a result of the analysis, it can be concluded that predicting an oil spill is difficult because of the complexity of the processes that occur during an oil spill, such as advection, spreading, evaporation, emulsification, dispersion, sedimentation, filtration into the ground, as well as changes in the composition and properties of oil. All these processes should be considered and described in a mathematical model to adequately assess the behavior of oil and the creation of oil spill response lan.

The choice of the model is determined by the task to be solved, in addition, each model needs to be adjusted or adapted to the specific conditions of the region in which the model will be used. The success in predicting the spread of oil at sea is determined not only by the ratios used to calculate the characteristics of a spill, but also by the ability to use the results of meteorological forecasts, marine dynamics predictions, including calculations of sea wind waves and ice cover characteristics supplied by models adapted and verified for the region in question. From the authors' point of view, a systematic approach to the task of predicting the spread of oil at sea is important. Information about the processes of oil transformation in the marine environment and models for their calculation, about the capabilities of remote sensing of the Earth for the detection and determination of characteristics of oil pollution at sea, discussion of problems associated with information support of the forecast and interpretation of its results seem equally important for the successful solution of the problem.

References

1. Reed M, Emilsen MH, Hetland B, Johansen O, Buffington S, Hoverstad B (2006) Numerical model for estimation of pipeline oil spill volumes. Environ Model Softw 21:178–189
2. Fay JA (1971) A physical processes in the spread of oil on a water surface. MIT. NTIS Report AD726281
3. Mackay D, Buist I, Mascarenhas R, Paterson S (1980) Oil spill processes and models. Report EE8. Environment Canada, Ottawa
4. Delvigne GAL, Sweeney CE (1988) Natural dispersion of oil. Oil Chem Pollut 4:281–310

5. Johansen O (1982) Dispersion of oil from drifting slicks. Spill technology newsletter, pp 134–149
6. Elliott AJ (1986) Shear diffusion and the spread of oil in the surface layers of the North Sea. Deutsche HydrographischeZeitschrift 39:113–137
7. Tkalich P, Chan ES (2002) Vertical mixing of oil droplets by breaking waves. Marine Pollut Bull 44(11):152–161
8. Zatsepa SN, Ivchenko AA, Solbakov VV, Stanova VV (2018) Prediction of the spread of petroleum products in the event of an emergency spill at sea: a scientific and methodological manual. JSC “Finpol”, Moscow, 140 p
9. Stanovoy VV, Eremina TR, Isaev AV, Neelov IA, Vankevich RE, Ryabchenko VA (2012) Modeling of oil spills in ice conditions in the Gulf of Finland on the basis of an operational forecasting system Oceanology 52:754–759
10. Stanovoy VV, Eremina TR, Karlin LN, Isaev AV, Neelov IA, Vankevich RE (2018) Operational and forecasting modeling of oil pollution spreading in the gulf of Finland. Scientific notes
11. Zhuravel VI (2012) Modeling of the behavior of possible oil spills during the operation of Prirazlomnaya ILSP: research report. NMC Risk Informatics, Moscow, p 87
12. Kucheiko AA, Zatyagalova VV (2010) Russian space technologies: new possibilities of operational monitoring and control. T-comm 2-20:18–21
13. Liungman O, Mattsson J, Massmann S (2013) Scientific documentation Seatrack Web. Physical processes, algorithms and references, 32 p
14. Ovsienko S, Zatsepa S, Ivchenko A (1999) Study and modeling of behavior and spreading of oil in cold water and in ice conditions. In: 15th international conference on port and ocean engineering under arctic conditions, Espoo, Finland
15. Ovsienko SN, Zatsepa SN, Ivchenko AA (2011) Mathematical modeling as an element of informational support for decision-making in choosing the strategy of marine environment protection from oil pollution. In: Proceedings of GOIN. M. Gidrometeoizdat, p. 213
16. Zatsepa SN, Ivchenko AA, Solbakov VV (2016) Methodical recommendations for predicting the spread of oil and petroleum products in the event of an emergency spill in offshore areas. Moscow, 101 p

Chapter 5
Marine Ecosystems and Features of Their Functioning

5.1 Marine Ecosystems

Marine ecosystems are those that have formed in the aquatic environment, with a dissolved salt content of about 35‰ ppm. These are mainly sodium and chlorine. Marine ecosystems cover almost 71% of our planet's surface and are part of the global ocean system and the Earth's hydrosphere structure.

Marine ecosystems are part of the biosphere, producing 32% of all net primary production. They can be divided into zones, depending on depth and coastline. Oceanic ecosystems have great depth and surface area. The open ocean is sparsely populated. It is inhabited mainly by whales, sharks and tunas, as well as benthic invertebrates [1]. The aquatic ecosystems near the shore are called coastal ecosystems. They are also referred to as coastal ecosystems: estuaries, salt marshes, coral reefs, lagoons, mangrove swamps.

Marine ecosystems have a significant impact on climate formation. Evaporation from their surface is the main source of water in the atmosphere, and currents are the temperature regulator.

Marine ecosystems, due to their great biological diversity, are resilient to many impacts. They successfully resist invasive species, natural pests and anthropogenic influences.

Marine ecosystems or freshwater, like terrestrial ecosystems, are built according to their inherent rules of formation. The main thing is that an ecosystem has as many species of living organisms as necessary to absorb and process the incoming energy of the Sun. Features of aquatic ecosystems are that they have internal complexity and non-linearity of connections, are subject to various external influences and are not closed, a large number of heterotrophic organisms and fast biotic circle, high stability, resistance and adaptability, population regulation is carried out by limiting resources or activity of predators. In addition, the world ocean ecosystem conserves significant amounts of excess carbon dioxide within itself. It is a global system with signs of continuity.

K. Pokazeev et al., *Pollution in the Black Sea*, Springer Oceanography,
https://doi.org/10.1007/978-3-030-61895-7_5

Geophysical processes have a huge impact on the functioning of marine organisms and their communities. Marine organisms, in turn, have a strong influence on their habitats, in particular by forming the chemical composition of seawater and bottom sediments. A true understanding of the laws of the ocean is therefore possible through an integrated study of the hydrophysical and biological processes occurring in the ocean.

There are several types of marine ecosystems, the most important of which are coral reef ecosystems, based on "symbiotic organisms"—corals consisting of intestinal-colontinental organisms—corals and algae in their bodies, these ecosystems have a rich biota; continental shelf ecosystems—the set of benthic organisms, including multicellular algae; the largest deep-water pelagic ecosystems are, firstly, organisms of the euphonic zone, including the smallest of plankton algae (up to 30) —coccolithophorides, which are at the limit of photosynthesis conditions (100–250 m); second, deep aphonic organisms feeding on detritus and dissolved organic matter, as well as many species of biotrophs, including deep-sea fish; finally, deep-sea benthos.

The coastal zone is a shallow line along the coast to the edge of the continental shelf with warm and nutrient-rich water. Its area is less than 10% of the ocean, but 90% of its biomass lives here. Locations on the coast where salt and fresh river water mixes are called estuaries. Here, the biomass is maximum and comparable to tropical forests. Coral reefs are also located in coastal areas of tropical and subtropical latitudes with water temperatures over 200 °C. Products in them are red and green algae. The consumer world is extremely diverse. One third of all marine fish species live here [2].

Marine organisms are an integral part of the marine ecosystem, in which all organisms are linked and depend on physical and chemical environmental conditions. Marine organisms, like all other living organisms, live by synthesizing organic compounds, exchanging matter and energy with the environment. Studying of substance and energy flows in marine ecosystem was the main ecological task of oceanology.

Marine organisms are characterized by a strong diversity in size, shape and way of life. The number of species living in the ocean is about an order of magnitude smaller than the number of species living on land. This is due to the great variety of environmental conditions on land. However, the ocean is home to species that do not exist on land, such as brown and red algae, echinoderms, bristle jaws and others.

5.2 Autotrophic and Heterotrophic Organisms of Marine Ecosystems

By type of energy source used, all marine organisms are divided into two types: autotrophic and heterotrophic.

Autotrophic *synthesize organic compounds* from inorganic ones. In the aquatic environment and using the energy of the sun's rays, they *produce oxygen* from carbon dioxide and increase their biomass. Autotrophic organisms do not eat other organisms, they consume inorganic compounds. The synthesis of carbohydrates and other compounds in autotrophic organisms is done through photosynthesis or chemosynthesis processes.

Rapid biomass growth does not always have a positive impact on ecosystem development and existence in general. By increasing their volume, plants can cut off light from deep into the reservoir, slow down internal nutrient metabolism and reduce oxygen content in the water. This will change the species composition of the ecosystem towards more chemosynthetic bacteria. These are microorganisms that feed on hydrogen sulfide. In the deep ocean, these bacteria are the source of nutrition for other living organisms. Like giant tube worms. In other water bodies, they don't find their consumers. So they quickly turn a pond into a swamp and then into peat deposits.

Only through these processes do inorganic substances get involved in the life cycles of marine organisms. Photosynthesis plays the main role, as it is used to create the bulk of organic compounds in the ocean. In the photosynthesis reaction carbon dioxide, water and the energy of sunlight create sugar, water and oxygen, energy in a chemical form is stored in a sugar molecule.

The reaction of photosynthesis can occur only in some types of marine organisms (algae, grasses, diatoms, some flagellates, etc.).

During chemosynthesis sugar is formed from carbon dioxide and hydrogen sulphide. The energy necessary for this reaction, is drawn from some chemical compounds, for example, hydrogen sulfide, in the course of their oxidation. On ocean floor about so-called "black" and "white" smokers there are surprising communities of sea organisms in which basis organic substances arising in the process of chemosynthesis lay. Chemotrophs can use not only hydrogen sulphide, but also nitrogen and sulphate. The study of marine communities based on chemotrophic processes is of considerable practical interest, as such systems can, in principle, be created on land artificially for the disposal of harmful substances.

In addition to photosynthesis and chemosynthesis, there are various processes of secondary synthesis in living organisms, during which larger and more complex organic substances are created.

Heterotrophs eat either live or dead tissue of other organisms. This organics provides chemical energy to heterotrophic organisms for secondary photosynthesis reactions.

Since most of the organic matter on the planet (and in the ocean) is formed by photosynthesis, it can be said that life on Earth is based on the energy of sunlight, which is retained by the biosphere through photosynthesis and is chemically transported as part of food along a trophic chain from one organism to another.

The division of marine organisms into autotrophic and heterotrophic is made by sources and methods of energy production. It is possible to divide marine organisms by their habitat. There are two main habitats for marine organisms: the seabed, or benthic zone, and water above the seabed, or pelagic zone. The organisms present in the benthic zone are called benthos. Benthos is subdivided into epibiotes inhabiting the bottom surface and endobiotes inhabiting the bottom sediments.

Organisms living in the pelagic zone are divided into necton (actively moving organisms) and plankton (passively moving organisms). Plankton is divided into phytoplankton (plant) and zooplankton (animal). This division is to a large extent conditional. For example, many species of marine animals change their lifestyle during their life cycle. Adults of some species may lead different lifestyles. For example, lobsters crawl on the surface of the sea floor and swim over it. Such animals are referred to as noncombenthic animals.

As already mentioned, the most important property of ecosystems is the transport of energy and matter. The solar energy absorbed by plants is transferred from them to animals and bacteria along the food chain as chemical energy. Animals and bacteria exchange carbon dioxide and mineral nutrients with plants. The flow of organic matter is closed in the sense that the same substances, which are part of the system and are replenished from the ocean, circulate between the system components.

The flow of energy is unidirectional, all incoming energy is dissipated, after all, as heat as a result of mechanical and chemical processes. It must be clarified that the biotic components of an ecosystem contain living and dead matter. Biogenic residues (organic detritus) make up a significant proportion of the total substance in the marine part of the biosphere. The main consumers of detritus are bottom organisms. Plants that form organisms as a result of photosynthesis are called primary producers, and the total amount of matter produced over a given period of time is called gross primary production. According to the method of nutrition, 5 types of organisms are distinguished: herbivorous; carnivorous (predatory); scavenger or corpse eaters; organic detritus eaters and, finally, decomposing dead organic substances (reducents) into minerals.

5.3 Trophic Levels of the Marine Ecosystem

In the process of nutrient regeneration, marine organisms turn complex organic substances into simpler ones, which are assimilated by plants from water. The biological interaction between organisms associated with food consumption is called trophic. Organic matter is transmitted through trophic levels, starting with the first, where phytoplankton forms the gross primary production.

The second trophic level is formed by vegetable eating zooplankton, the production of zooplankton is called secondary. Rank 1 predators eating zooplankton form the next trophic level and create an organic substance called tertiary production. This is followed by predators of higher ranks. The nutrient system is closed by the processes of bacterial decomposition (indirect regeneration) and metabolic processes occurring in plants and animals (direct regeneration).

The described scheme of trophic levels is approximate. More precisely, it is more correct to speak not about the food chain but about the food network. In this case there are several different organisms at different trophic levels, and metabolism occurs between organisms of different levels. In this case there are cross trophic connections, a trophic network is formed. The ecological efficiency of the trophic level is determined by the amount of energy transferred to the next trophic level. In general, the amount of production coming to any level is determined by the production of primary products for the ecological efficiency of all previous levels. Ecological efficiency values of trophic levels are rather low, about 0.1, so relatively little energy is transferred to the higher trophic levels.

In the marine ecosystem, the spatial organization of energy and substance flows has an important feature. The zone in which photosynthesis occurs is located at the surface of the ocean and there is no photosynthesis in the deep layers. In these layers of lack of photosynthesis, which are separated from the photosynthetic zone by a considerable distance, energy is delivered in a chemical form. The transfer of energy to the deep layers occurs along the food chain and due to the settling of organic detritus.

Thus, there is a flow of nutrients from the surface layers, directed deep into the ocean. If there were no opposite processes, the upper ocean would rapidly lose its nutrients and life in the ocean would cease. Reverse nutrient inputs to the surface waters of the ocean occur as a result of several physical processes, primarily upwelling, equatorial water uplift, high latitude water uplift, and seasonal thermal wedge collapse due to convection. In some coastal areas of the world's oceans, the intensity of upwelling is particularly high. The high concentration of nutrients creates favorable conditions for the development of phytoplankton and other organisms. These areas of the world's oceans are the most productive.

5.4 The Productivity of Marine Ecosystems

The parameters of marine ecosystems change significantly in space and time. The established stock, or biomass, determines the biomass value of certain specific organisms or the total biomass of all organisms. Depending on the research objectives, the established stock is expressed as the number of individuals per unit area (or volume) or as mass per unit area (or volume).

The intensity of solar radiation is determined by geographic and meteorological conditions. With depth, the intensity of solar radiation falls very rapidly. At a depth of 10 m 10% of the energy entering the water surface is available for photosynthesis. At a depth of 100 m only 1% of the energy is available for photosynthesis. The photosynthesis zone is therefore limited to several tens of meters. In offshore waters, where there is more suspended matter than in the open ocean, the absorption of light in the upper layer of water is even greater and the thickness of the photosynthetic layer even less. Water temperature also has a strong influence on primary production. The photosynthesis rate is maximum (all other conditions being equal) in a given temperature range. For most areas of the ocean, water temperature is below this optimum for many marine organisms. Therefore, the seasonal heating of the water leads to an increase in photosynthesis rate.

Besides water and carbon dioxide, plant development requires nutrients (biogens). In the ocean, basic nutrients include nitrogen, phosphorus and silicon. Nitrogen and phosphorus are the elements needed to build cells. Phosphorus is involved in energy metabolic processes. In plants, the ratio of "nitrogen to phosphorus" is about 16:1. In some coastal areas, phosphorus acts as a limiting photosynthesis element. Two groups of phytoplankton organisms, diatomeias and dinoflagellates, consume a lot of phosphorus. The rapid removal of silicon from the water when they build their skeletons creates a shortage of silicon and limits their development.

Emission of phytoplankton by zooplankton affects the value of primary production. Therefore, the intensity of ejection is one of the most important factors determining the value of primary production.

The productivity of seawater varies greatly in space. High productivity is typical for shelf waters, upwelling zones, where intensive enrichment of surface waters with nutrients occurs. Areas of intense coastal upwelling off the coast of Peru, Oregon, Senegal and southwest Africa are very productive.

High productivity has been noted in the areas of boundary currents, where good mixing of water masses takes place. The high productivity of the shelves is due to the fact that the water in them contains relatively many nutrients due to river runoff, nutrient inputs from bottom sediments (due to the relative proximity of the bottom), better lighting and relative warmth of the water masses. In the open ocean, the area of high productivity is small as vast areas of the ocean are occupied by anticyclonic cycles in which surface water sinking occurs.

In high latitudes (above 50°), a seasonal thermal wedge with convective mixing of water masses is destroyed. In circumpolar regions of the ocean there is an upward movement of deep masses. Therefore, these ocean latitudes belong to highly productive areas. As we move further towards the poles, productivity begins to drop due to lower water temperatures and lower illumination. The ocean is characterized not only by spatial variability in productivity but also by widespread seasonal variability. Seasonal variability in productivity is largely due to the response of phytoplankton to seasonal changes in habitat conditions, primarily in lightness and temperature. The greatest seasonal contrast is observed in the temperate zone of the ocean.

Let's give some information about the food that people get from the ocean. The ocean covers about 71% of the planet's surface area, receives much more solar energy than land, but the ocean contributes about 1% to human nutrition. On the other hand, the ocean is an important source of animal proteins—about 5–10% of the protein consumed comes from the ocean. Many areas of the ocean have the same total annual primary production as the land. However, the overall low value of ocean production is due to the fact that humans feed on marine organisms at higher trophic levels. Fishery-rich areas coincide with areas of the ocean with high primary productivity. Three areas of primary productivity in the ocean can be identified: the open ocean zone, the coastal zone, and the upwelling zone. The upwelling zones are about six times more productive than the open ocean zone. The coastal zone is an intermediate productivity zone. Since the upwelling and coastal zones are much smaller than the open ocean zone, their total productivity is smaller than the upwelling and coastal zones. However, the number of trophic levels between the level of primary producers and the level of commercial fish and crustaceans in these zones varies significantly. While in the open ocean, food chains ending in capture fisheries have on average 5 trophic levels, in the upwelling zone of such levels one to two. This explains the high productivity of upwelling zones. Coastal zones, which are much larger than upwelling zones, have approximately equal fish productivity. This is because there are more trophic levels between primary producers and commercial fishes in the coastal zone than in upwelling zones.

Biological contamination is the introduction of alien species of animals, plants and microorganisms into biocenosis. The main sources of this process are: water transport, fisheries, industrial discharges and sewage, as well as targeted anthropogenic activities for the introduction of certain species of fish, mammals, etc. The largest scale in recent decades has been the transfer of alien fauna to other regions of the world ocean with ballast waters of large vessels.

Let's give one example of ocean biological pollution. In 1987 in the Black Sea were found individual instances of mnemiopsis comb. The mnemiopsis comb lives in lagoons on the Atlantic coast of the USA. Its numbers are limited in these natural conditions. It was probably accidentally brought into the Black Sea with ballast waters of ships. In the Black Sea, he has no natural enemies like those found in lagoons on the Atlantic coast. Grebnevik found himself on the same trophic level as aurelian jellyfish and won the competition. For 2–3 years, the density of jellyfishes in the Black Sea has decreased by almost 10 times, and the average size of jellyfishes decreased by about 4–6 times. The total mass of mnemiopsis reached 900 million tons, which is 10 times more than the entire annual catch in the sea [3].

Due to the development of the ridge wheel, the significant changes in the structureof meso-macroplankton communities have been observed. First of all, its settlement led to a sharp decrease in the biomass of food for mnemiopsis and fish of high-calorie zooplankton. Due to this, the comb appeared as a serious trophic competitor of commercial plankton fish - hams, horse mackerel and tulip. The total catch of fish in the system of Marmara-Black-Azov Sea, which reached in 1985–1986. 856–906 thousand tons in 1989 fell to 640 thousand tons. Khamsa catch in the northern part of the Black and Azov Seas in 1980–1988 ranged from 240 to 126 thousand tons. Catch of Khamsa in the northern part of the Black and Azov Seas in 1980–1988 fluctuated from 240 to 126 thousand tons, and in 1989 it decreased to 70 thousand tons, catch of horse mackerel fell from 110 to 115 thousand tons to 3 thousand tons, catch of tulle in 1970–1987 was 77–130 thousand tons, and in 1988–1989 it decreased to 36–40 thousand tons. Obviously, the observed catastrophic decrease in catches was directly related to undermining mnemiopsis of the feeding base and, possibly, direct eating of fish larvae during spawning [4].

This situation continued until 1999. In 1999, a new crest was invaded into the Black Sea—a beroy. Presumably, the beroy comb got into the Black Sea with warm water from the Mediterranean basin or was brought in like a mnemiopsis with ballast water. The beroe ridge feeds on the mnemiopsis ridge. The introduction and development of beroy has led to a sharp decrease in the biomass of mnemiopsis and, as a result, to the growth of zooplankton and fish larvae, and later of pelagic fish stocks. The mass development of berries in the Black Sea resulted in a decrease in the trophic press of mnemiopsis for plankton community. The paired existence of ridges in other sea basins of the world ocean provides a natural balance of their numbers. For example, off the coast of the Americas, the Atlantic Ocean is home to Mnemiopsis leidyi and Beroe cucumis, which destroy the first species. Thus, in the Black Sea, after crest invasion, Beroe cucumis develops a situation similar to that existing in other ocean regions. Let's consider a mathematical model of this system, corresponding type of pairwise interspecies interaction - competition [5].

Let the total mass of jellyfish M_1, and the mass of comb M_2. Total food supply M_0. Then the system of equations describing the change in masses of M_1 and M_2 can be written in the form:

$$\begin{aligned} \frac{dM_1}{dt} &= \gamma_1 M_1 M_0 - \varepsilon_1 M_1, \\ \frac{dM_2}{dt} &= \gamma_2 M_2 M_0 - \varepsilon_2 M_2 \end{aligned} \tag{5.1}$$

Members $\varepsilon_1 M_1$ and $\varepsilon_2 M_2$ describe the loss of species M_1 and M_2, respectively. Let's divide the first equation into $\gamma_1 M_1 M_0$, the second equation into $\gamma_2 M_2 M_2 \gamma$, subtract the second equation from the first and integrate it. As a result we will get:

$$\frac{M_1^{\frac{1}{\gamma_1}}}{M_2^{\frac{1}{\gamma_2}}} = const \exp\left[\left(\frac{\varepsilon_2}{\gamma_2} - \frac{\varepsilon_1}{\gamma_1}\right)\right] \tag{5.2}$$

If $\frac{\varepsilon_2}{\gamma_2} > \frac{\varepsilon_1}{\gamma_1}$, then M_1 will grow, M_2 will shrin. If $\frac{\varepsilon_1}{\gamma_1} > \frac{\varepsilon_2}{\gamma_2}$, then as time grows the mass of M_2 will prevail and M_1 will tend to zero.

This last case is realized in the system of aurelium jellyfish—mnemiopsis comb. The comb, as having a lower mortality rate and consuming more food, has almost completely replaced jellyfish with aurelium. At the same time, we have received the famous Voltaire's theory: if species live in one ecological niche, some of them always crowd out the others.

Small living creatures such as plankton in the oceans at the beginning of the food chain absorb chemicals during their lifetime. Since plankton and other small creatures are quite resistant to destruction, their bodies accumulate chemicals in higher concentrations than the surrounding water or soil.

These organisms, in turn, absorb other small animals and the concentration of toxic substances increases again. These animals are then eaten by large animals that can travel long distances with even higher concentrations of chemicals inside their bodies. Animals that are higher up the food chain, such as seals, can have pollution levels millions of times higher than the environment. And polar bears that feed on seals can have pollution levels 3 billion times higher than their environment.

Another factor of biological pollution is the deliberate introduction of "useful" species of fauna by humans. Here such actions as settling in the water area of the Black and Azov Seas of Pilengas, and in the Barents Sea—pink salmon and Kamchatka crab can be cited as examples. These species have occupied their ecological niches, quickly multiplied and gained commercial importance without causing any visible damage to biocenosis. However, together with relatively large animals, various parasites, microorganisms, etc. are inevitably transported. As a result, they can move to other species in a new location and can cause some biota damage in the future. All these phenomena have not yet been studied well enough.

In conclusion, we note that the introduction of alien organisms (invasion) is a serious environmental problem of global scale. Moreover, it is in recent years that there has been a sharp increase in the transfer of various non-typical species for biocoenosis. First of all, it is caused by the increase in the volume of transportation of large-capacity vessels (especially supertankers). Increasing aquaculture production is also a serious factor.

5.5 Conclusions

The peculiarities of functioning of marine ecosystems and their differences from fresh water are considered.

The main types of marine ecosystems are analyzed in terms of depth of location and energy used.

Explains the reasons for the different biological productivity of the ecosystems of the World Ocean, depending on the geographical location.

Case studies show that invasion of marine ecosystems by alien species has now become a serious environmental problem of global concern.

References

1. http://ecology-of.ru/eko-razdel/ekosistemy-sreda-obitaniya-voda
2. https://bstudy.net/740470/estestvoznanie/morskie_ekosistemy
3. Grishin AN, Mikhneva VV, Khineva EN (2006) How does a mnemiopsis comb survive in the Black Sea?. Fishery of Ukraine 3(4):3–7
4. Gubanov EP (2006) Ecological aspects of the Black Sea bioresources state (in Russian). In: Gubanov EP (ed) Modern problems of the Azov-Black Sea basin ecology. Proceedings of the II International Conference, UGNIRO, Kerch, pp 10–16
5. Menshutkin VV, Pokazeev KV, Filatov NN (2004) Hydrophysics and ecology of lakes, vol II, Ecology. Physics Faculty of Moscow State University, Moscow, 280 p

Chapter 6
General Oceanographic Characteristics of the Black Sea

6.1 Introduction

The Black Sea is the world's largest distributed (brackish water) sea basin, which is due to the narrow and shallow straits, one of the most isolated seas from the World Ocean. The uniqueness of this semi-isolated body of water from the World Ocean lies in a significant amount of fresh water flow into it, 70% of which comes from industrial areas of Europe, bringing with it a significant amount of various pollutants.

The area occupied by the Black Sea is 420,325 km^2, the average depth is 1271 m, the greatest is 2210 m. The total volume of water is 547,015 km^3. The volume of water contaminated with hydrogen sulfide is 475,000 km^3. The deep water mass of the Black Sea with temperature 8–90 °C and salinity 22.2–22.4‰ has the volume of 228,680 km^3. The salinity of the surface waters of the sea is low (5–10‰) in the estuarine areas and is about 18‰ in the sea centre. The average vertical transfer rate in the Black Sea is about 10^{-4} cm/s. However, in local areas it may be 10^{-3}–10^{-2} m/s, which should certainly be taken into account when dealing with issues of protecting the marine environment from pollution.

According to modern estimates, the Black Sea is one of the most polluted areas of the world's oceans. Although it belongs to the Atlantic Ocean basin, the Black Sea is an inland sea, located at the junction of Europe and Asia. The whole territory around the sea, as well as the rivers flowing into it, since ancient times is densely populated by people. The sea washes the coasts of Russia, Ukraine, Romania, Bulgaria, Turkey, Georgia and Abkhazia. The largest rivers in Europe: the Danube, the Dnieper, the Dniester, and more than 300 other rivers flow into the Black Sea. During the year, the rivers bring more than 346 km^3 of fresh water into the sea, with 80% of this water being discharged to the north-western shelf part, mainly the Danube and Dnieper. The catchment area of the Black Sea basin is shown on the map (Fig. 6.1).

K. Pokazeev et al., *Pollution in the Black Sea*, Springer Oceanography,
https://doi.org/10.1007/978-3-030-61895-7_6

Fig. 6.1 Catchment area of the Black Sea basin https://www.welt-atlas.de/map_of_black_sea_1-1033

The study of migration and residence time of contaminants in the Black Sea allows estimating the ability of the sea to self-cleaning and is an actual scientific problem [1].

The Black Sea deserves to be nicknamed the Bacterial Sea. About 80% of its water column and bottom sediments within the anaerobic zone are deprived of other life forms than bacteria. Surrounded by industrially and agriculturally developed countries, the Black Sea serves as a natural sedimentation tank. The ratio of catchment land area to its mirror area (422,000 km^2), equal to 4.4, is an order of magnitude higher than the average for the ocean—0.4. The rivers carry into the sea poisonous chemicals from fields, pesticides, detergents, saturated and aromatic hydrocarbons, heavy metals salts and other toxic substances. It receives a large amount of fertilizers washed off the fields and organic substances, leading to over-fertilization—eutrophication of the pond.

In addition to river runoff, the increasing stress on coastal areas is caused by stormwater, domestic and industrial runoff from enterprises, communities and recreational complexes located on its banks. Their local impact is often catastrophic. About 180 thousand vessels with the total displacement of 125 million tons pass through the straits annually, with more than 12 thousand tons of oil entering the sea.

Strong impacts on the biological communities of the sea and its fishing potentials are caused by irrational, excessive fishing, and in recent years by the shift of new settlers from other areas of the world's oceans, which is destroying the autochthonous communities of the Black Sea.

6.2 General Oceanographic Characterization of the Black Sea

The Black Sea is one of the most studied seas of the World Ocean, there have been numerous marine expeditions, hundreds of scientific articles and dozens of monographs of various directions are devoted to it. General features of the hydrological structure of the Black Sea have been known since the first half of the XX century by the works of F. F. Wrangel, I. B. Spindler, N. M. Knipovich; further they have been constantly specified, in particular, in the generalizing works of A. K. Leonov, D. M. Filippov, A. S. Blatov, V. S. Tuzhilkin (1968, 1968, 2008, 2008, 2008).

In this paper, information on the circulation and thermochaline structure of Black Sea waters is given and used to the extent necessary to characterize the transport routes of natural and anthropogenic pollutants. The main source of such data is instrumental measurements of the velocity vector of sea currents and the spatial thermohaline structure of sea waters [2].

The state of the Black Sea ecosystem differs significantly in its coastal and open areas, due to the peculiarities of circulation. The principal scheme of average Black Sea water circulation has been known since the beginning of the 20th century: along the outer edge of the continental slope the deep part of the sea rounds off, moving counter-clockwise the flow of the Main Black Sea Current (MBSc). Two weaker cyclonic cycles are inscribed into the western and eastern extensions of the deep water area. The formation of the Black Sea salinity field is influenced by two climate-forming factors: the spreading influence of river runoff and the intensity of salt exchange between the Black Sea and Marmara Sea through the Bosporus Strait [3].

The cyclonic Main Black Sea current (MBSc) is traced throughout the sea, creating extensive cyclonic cycles in the central parts of the western and eastern regions (Fig. 6.2) and a number of smaller cyclonic and anticyclonic vortices. In the eastern part of the sea to the east of the main sea, there are also cyclonic cyclones, while in the southeast part of Batumi, there are intense anticyclonic cyclones. Cyclonic vortexes ascend and sink in the anticyclonic and external periphery of the main Black Sea current.

As a result of meandering of the main Black Sea current (MBSc) and its lateral friction against the mainland slope, an anticyclonic vortex of the current in the coastal zone near the shelf edge, where anticyclonic vortexes are generated. In the anticyclonic vortexes, water accumulation and subsequent sinking occur, which coincides with sinking of water at the periphery of cyclonic cycles. Their addition causes active water immersion near the continental slope and the appearance of a

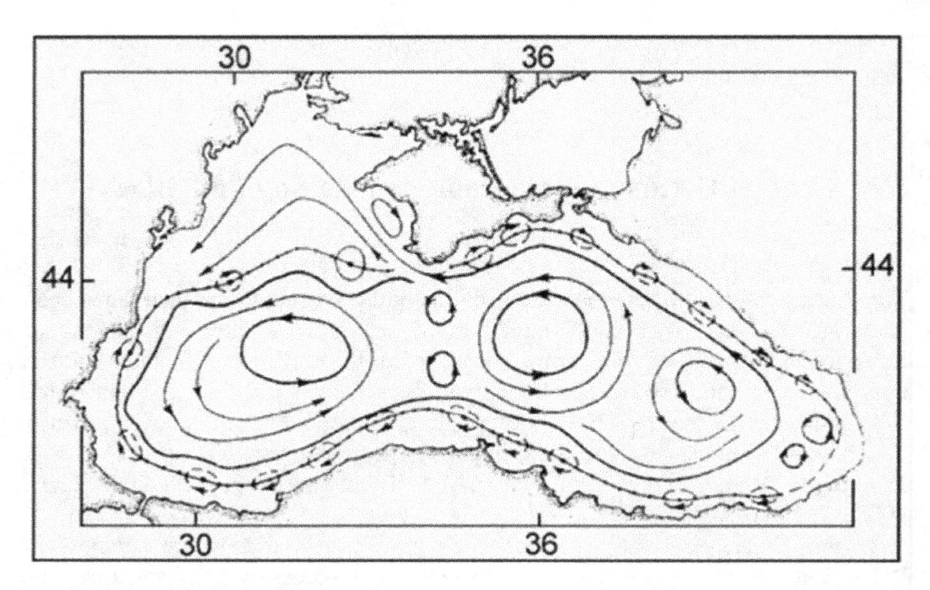

Fig. 6.2 General scheme of the Black Sea surface water circulation [23]

quasi-stationary convergence zone. In calm weather, convergence on the outer boundary of the main current is clearly manifested on the surface by slicks and floating debris accumulations. Off the coast of the Caucasus and Anatolia, the convergence zone is 5–7 miles away from the shore, and in the west and especially in the northwest, along with the edge of the shelf, it goes far away and is expressed weaker.

The existence of a convergence zone beyond the edge of the shelf predetermines the compensatory rise of waters on the continental slope, which is amplified when anticyclonic vortices pass along the coast. The water rising from the cold intermediate layer contributes to a sharp drop in temperature along the coast. A compensatory rise of water may also occur in the convergence zone. Thus, a frontal zone similar to the sloping fronts of the open ocean is formed near the Black Sea continental slope.

This zone serves as a powerful hydrochemical barrier with maximum concentrations of Zn, Cr, Cu, Pb, Hg, Cd, organochlorine pesticides and polyaromatic hydrocarbons. When heated and distributed coastal waters are mixed with colder and saltier waters of the main Black Sea current, water compaction and sinking occur. Gravitational settling of organic and inorganic suspension as well as turbulent isopic movements result in suspension enrichment of waters along the mainland slope. Convergence of coastal waters takes pollutants from the surface layers to a depth and to some extent isolates the open waters of the sea from the main volume of coastal anthropogenic runoff. These contaminants are carried by deep water through the sea.

The results of systematic long-term expeditionary studies of the processes of autumn-winter convective mixing in the northwestern part of the Black Sea were summarized in the monograph of the Department of Oceanology of the Faculty of Geography of Moscow State University [4].

I. M. Ovchinnikov and Y. I. Popov suggested the hypothesis of the cold intermediate layer in the Black Sea (CIL) formation. According to this hypothesis [5, 6], (CIL) waters are formed in the centers of major cyclonic cycles, similar to the previously open processes of deep convection in the Greenland Sea and the Gulf of Lyon in the Mediterranean Sea.

An important generalization work was the monograph devoted to the study of the variability in the hydrophysical fields of the Black Sea [7]. It deals with the main types and mechanisms of spatial and temporal variability of hydrophysical fields, including synoptic and mesoscale ones. The climatic thermohaline fields were calculated for 25 thousand stations, and the hydrological structure, CPS formation and transformation, and geostrophic circulation were analyzed.

The decade of the 1990s was characterized by increased international cooperation in the study of the Black Sea, including joint maritime expeditions. This partly compensated for the division of the sea area into national economic zones, which had by then been formalized and were subject to the permitting character of oceanographic surveys. The Convention for the Protection of the Black Sea from Pollution was signed in 1992 the CoMSBlack programme for joint scientific research on the Black Sea was launched in 1991, and the Black Sea Environmental Programme (BSEP) has been implemented under the auspices of the Global Environment Facility (GEF) since 1993.

On the basis of the increased volume of available oceanographic and meteorological information obtained, including new, modern measuring instruments, in the last decade [8] the climatic variability of the hydrological regime in the Black Sea in a century time interval was considered. According to [8], the main reason for regional differences in climatic changes in the Black Sea from other areas of the World Ocean is the two-layer hydrological structure of waters and the inland position of the sea, determining the degree of response of the basin to external influences.

According to the work data [8] as a whole by sea, the positive balance of fresh water in the layer 0–50 m is compensated by the inflow of salts from deep layers, the components of the salt balance are shown in Fig. 6.3.

Among the water masses of the Black Sea, the greatest amplitude of seasonal fluctuations in water volume is the coastal water mass. Its volume (at criterion of allocation $S < 17.8‰$) increases from winter to summer by 3 times, with a maximum in July, by the time of change of water exchange direction between the shelf and the central part of the sea.

By the end of the 1990s, the volume of expeditionary research had drastically decreased, and synoptic surveys of the entire basin had not been conducted since then. Prof. Vodyanitsky (IAEA programme), Knorr in 2001, 2003, Endeavor in 2005, and M. Vodyanitsky in 2005. S. Merian in 2013. Regular surveys of NIS "Akvanavt" (S&A RAS) were carried out in the Russian zone, NIS "Akademik"

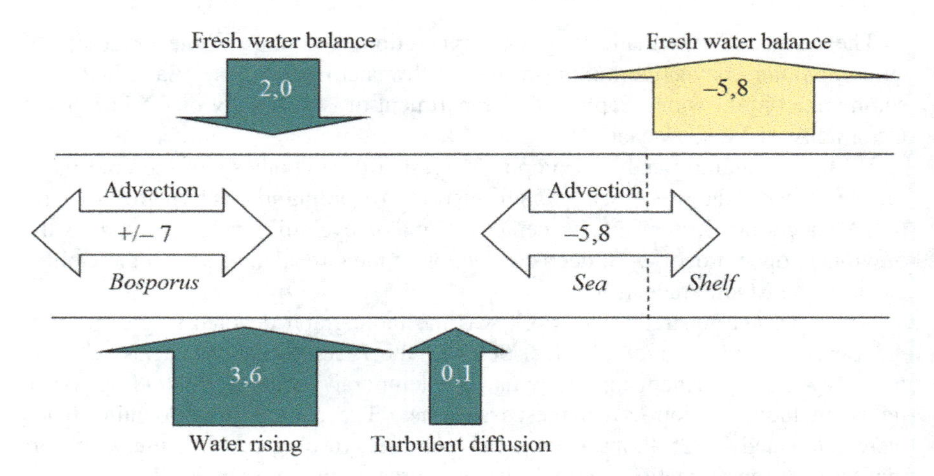

Fig. 6.3 Components of salt balance in the layer 0–50 m of the Black Sea (billion tons per year) [8]

(S&A BAN)—in the Bulgarian sector, NIS "Prof. Vodyanitsky"—in the Ukrainian zone. Vodyanitsky" (INBYUM) and MGI small-capacity vessels, since 2014. NIS "Prof. Vodyanitsky". Vodyanitsky" conducts oceanographic expeditions of IMBI RAS and MGI RAS in the Russian economic zone.

In recent decades, thanks to the application of high-precision hydrophysical probes, new important oceanographic results have been obtained with respect to the vertical structure of deep and near-bottom waters of the Black Sea [9]. On the Hydrophysical and Hydrochemical Homogeneity of the Black Sea Deep Waters [9–11]. In particular, a quasi-isothermal layer was found at depths of 500–700 m, theoretical assessments of the existence of a 300–400 m thick bottom boundary layer were confirmed, and a developed thin structure in the deep layers was recorded. In [12, 13] the origin of the quasi-thermal layer is explained by the compensation of two heat flows: the advection of marble waters and the geothermal flow. The high spatial homogeneity of temperature and salinity of the deep layers was confirmed in [14].

In the Bosporus Strait and the Pribosporus area, mainly by Turkish oceanologists, a large number of studies have been carried out to better understand the processes of formation of deep waters in the Black Sea. It was confirmed that the waters of the Marble Sea after leaving the Strait follow the S-trajectory following the relief of the Bosporus Canyon bottom.

As the number of in situ measurements declined, satellite-based remote-sensing and drifting technologies began to play an important role.

Since 2002, with the launch of ARGO's drifting profiling floats in the Black Sea, it has become possible to measure temperature and salinity to great depths. In [15] considered the results of ARGO buoys in the application to the issues of circulation in deep layers. The spatial and temporal variability of the thermochalinic structure of the upper 300 m layer of the Black Sea using the method of decomposition by empirical orthogonal functions according to the data of profiling floats for the period 2002–2012 was considered in [16].

It is concluded [8] that during its geological history, the Black Sea basin, depending on its connection with the World Ocean, passed through various stages of the Khalini regime—brackish: 5–13‰ (Chaudian, Euxian basins), marine: 15–20‰ (caradenism, Uzunlar, Black Sea basins), ocean: up to 30‰ (Qarangat basin). The positive trend of sea salinity in the modern period may indicate that the process of basin salinization after the end of the last ice period has not yet been completed. The main reason for the regional differences in climatic changes in the Black Sea from other areas of the World Ocean is the inland position of the sea, which determines its isolation, weak external water exchange, two-layer hydrological structure of waters and increased response of the basin to atmospheric impacts.

According to [8] The character of long-term variability of thermohaline fields in the upper 100-m layer of the sea with predominance of interdecadal fluctuations sharply differs from the tendencies in the layer of permanent pycnocline and deep layers, where the inflow of marble sea waters causes weak but steady heating and salinization. The leading role in low-frequency variability of the sea heat reserve is played by the intensity of winter convection.

The differences in the characteristics of seasonal and inter-annual temperature and salinity variability in the Black Sea are related to changes in the overall intensity and redistribution of the relative role of the thermal and water balance components.

Although the Black Sea has been extensively explored, after more than 100 years of research, many questions remain open. They include: peculiarities of processes of winter convective mixing in different climatic conditions, stability of seasonal course of thermohaline water structure on the inter-decadal scale, reasons of long periodical tendencies in constant pycnocline, characteristics of water, salt balance of the sea and water exchange through the Bosporus, the degree of influence of large-scale atmospheric circulation on long-term changes in the hydrological regime of the sea, as well as sea pollution levels, which have changed significantly in recent years.

The problems, the study of which has not lost relevance at the present time, should include estimates of sea pollution levels, especially its shallow and shelf areas in terms of studying their impact on the ecological state of the deep sea ecosystem, taking into account the specifics of the hydrodynamic regime of the sea, which will be considered in the next parts of the book. In this direction, the number of publications is very limited, especially in the last decade, it should be noted the work on the impact of the Black Sea and Karkinit Gulf on the state of the ecosystem of the deep sea, especially the Northwest shelf of the Black Sea (NWSBS) layer [17, 18]. Among earlier works in the above direction it is necessary to note works [19–22, 25–27].

6.3 Conclusions

The main historical stages of the study of the Black Sea as one of the most isolated seas from the world ocean are given.

The uniqueness of the hydrology of the sea is considered, due to the presence of a pronounced vertical density stratification of waters, leading to a sharp stratification of the sea into a thin desalinated upper layer and a deep salt layer.

Modern ideas about peculiarities and specificity of water circulation in the sea are given.

The possibilities of modern methods and techniques for refining and improving assessments of the main oceanographic characteristics of the Black Sea are shown.

References

1. Ivanov VA, Pokazeev KV, Sovga EE (2006) World ocean pollution. Moscow State University, 163 p
2. Hydrometeorology and hydrochemistry of the seas of the USSR (1991) T. IV Black Sea. Exhibit 1. Hydrometeorological conditions. Pod of editorial board of A.I. Simonov, E.N. Altman.-Sant P-g: Hydrometeoizdat, 430 p
3. Latif MA, Qzsoy E, Oquz T, Unluata U (1991) Observations of mediterranean inflow into the Black Sea. Deep-Sea Res 38(Suppl.2):711–723
4. Dobrovolsky AD (1977) Convective stirring in the sea. Moscow State University, 239 p
5. Ovchinnikov IM, Popov YI (1987) Formation of the cold intermediate layer in the Black Sea. Oceanology T. 27, № 5:739–746
6. Ovchinnikov IM (1990) Peculiarities of the cold intermediate layer formation in the Black Sea under the extreme winter conditions (in Russian). GOIN (in russian), 990. Exhibit. 190, 132–151
7. Blatov AS (1984) Variability of hydrophysical fields of the Black Sea. L.: Gidrometeoizdat, 239 p
8. Belokopytov VN (2017) Climatic changes in the hydrological regime of the Black Sea. Abstract of the dissertation for the degree of Doctor of Geographical Sciences in specialty, 25.00.28—Oceanology, Sevastopol. MGI RAS, 42 p
9. Kushnir VM, Lebedeva TP, Linskaya EB (2000) Layered structure of the bottom boundary layer in the Black Sea. Morskoy Hydrophys J № 6:45–55
10. Volkov II (2002) In: Zatsepin AG, Flint MV (eds) Complex research of the north-eastern part of the Black Sea. Nauka, Moscow, pp 161–169
11. Samodurov AS (2009) Pridonny boundary layer in the Black Sea: formation of the stationary state. Mar Hydrophys J № 1:16–25
12. Samodurov AS, Ivanov LI (2002) Balance model for calculation of the average vertical flows of liquid, heat, salt and dissolved chemical substances in the Black Sea thermohalocline. Mar Hydrophys J № 1:7–24
13. Ivanov LI, Samodurov AS (2001) The role of lateral fluxes in ventilation of the Black Sea. J Mar Syst 31(1–3):159–174
14. Polonskiy AB, Lovenkova EA (2006) Long-term tendencies in the Black Sea deep thermochalinic characteristics variability. Mar Hydrophys J № 4:18–30
15. Korotaev GK, Oguz T, Riser S (2006) Intermediate and deep currents of the Black Sea obtained from autonomous profiling floats. Deep-Sea Res II 53:1901–1910

16. Ivanov VA et al (2013) Use of the ARGO float thermohaline data in numerical model validation and statistical analysis in the Black Sea. In: International conference Marine Research Horizon 2020, p 86, Varna, Bulgaria, 17–20 Sept 2013
17. Sovga EE, Kirilenko NF (2014) Long-term dynamics of biogenic elements in the ecosystem of the north-western shelf of the Black Sea (in Russian). Mechnikov OGNU Vestnik 1 (20):105–112
18. Ivanov VA, Sovga EE, Khmara TV, Zima VV (2018) Thermohalinic regime of the Karkinit Gulf water area and ecological consequences of natural use. J Ecol Saf Coast Shelf Sea Zones №3:22–33
19. Sovga EE (2002) Features of mechanisms of functioning of the Black Sea ecosystems of the shelf and pelagiali. The abstract of the dissertation for a scientific degree of doctor of geographical sciences. Sevastopol, MGI NAS of Ukraine, 37 p
20. Eremeev VN, Latun VS, Sovga EE (2001) Influence of the anthropogenic pollutants and ways of their transfer on the ecological situation in the north-western area of the Black Sea. Mar Hydrophys J № 5:41–55
21. Brass VS (1995a) Influence of anticyclonic vortices on the water exchange between the north-western shallow water and the deep water part of the Black Sea. In: Black Sea integrated environmental studies. MGI NAS of Ukraine, Sevastopol, pp 37–47
22. Brass VS (1995b) Influence of the water circulation near the Crimean coast on the biologically active substances transfer and water exchange between the north-western and deep-water parts of the Black Sea (in Russian). In: Problems of Ecology and Recreation of the Azov-Black Sea Region. Simferopol, pp 163–165
23. Vinogradov ME, Shoemaker VV, Shushkina EA (1992) Black sea ecosystem. M. Nauka, 112 p
24. https://greenologia.ru/eko-problemy/gidrosfera/chernoe-more.html
25. Filippov DM (1968) Black Sea water circulation and structure (in Russian). Science:136 p (Moscow)
26. Tuzhilkin VS (2008) Seasonal and perennial variability of the thermochaline structure of waters of the black and caspian seas and processes of its formation: Dr. Geogr Sci 11.00.08. Valentin Sergeevich Tuzhilkin, 313 p
27. Korotaev GK, Oguz T, Riser S (2001) Ventilation of Black Sea pycnocline by the Mediterranean plume. J Mar Syst 31:77–97

Chapter 7
Hydrophysical and Chemical Features of the Black Sea and Their Influence on Pollutant Transport and Transformation in the Sea

7.1 Introduction

The main factor determining the hydrophysical and hydrochemical structure of the sea is water exchange through the Bosporus Strait. The higher (approximately 35 cm) level of the Black Sea compared to the Mediterranean results in the presence of waste water from the Black Sea, which carries surface slightly saline waters. In the bottom layers, due to the difference in water density, a compensatory flow from the Sea of Marmara develops, carrying water with a salinity of 36‰. The volume of transport by each flow varies from year to year and from season to season. On average, about 370 km^3 are carried out from the Black Sea through the straits and 180 km^3 are brought in each year, which is about 0.04% of its volume. Saltwater falls into the sea basin and maintains a dynamic equilibrium on the pycnocline that separates the surface waters distributed by the river runoff from deep salt water. The volume of surface water with reduced salinity does not exceed 10% of the total volume of water in the sea. In general, the separation of the upper active layer from deep water is determined by a salinity difference of 2.5–3.5‰.

Deeper pycnocline water density rises very slowly, which makes deep stratification unstable and facilitates convective exchange between waters of different depths. At a depth of more than 1 km, salinity is 22.31–22.36‰, with a relative density of 17.25–17.28 relative units. The water stability coefficient ($E10^{-5}$) is only 0.006. Low resistance of deep waters to the processes of winter mixing over the pycnocline domes in cyclonic cycles leads to the fact that mixing of the whole mass of Black Sea waters is fast and bottom waters reach the surface in 100–130 years, testifying to the intensive dynamics of processes at the border of aerobic and anaerobic sea zones.

The existence of stationary pycnocline, which prevents mixing of upper and deep waters, causes the formation of extremely sharp oxycline (drops in oxygen

K. Pokazeev et al., *Pollution in the Black Sea*, Springer Oceanography,
https://doi.org/10.1007/978-3-030-61895-7_7

concentrations from 5 to 7 ml/l, typical for mixed surface and cold intermediate layers, to 0.35–0.5 ml/l).

The gradient of oxygen concentration in it can reach 0.2–0.55 ml/l and higher values. Oxycline is formed by oxidation of the dropping dead organic suspension, which is trapped on pycnocline, and due to oxidation of rising from the depth of the reduced forms of iron, manganese, nitrogen. About 30% is spent on oxidation of organic matter, and about 70% of the total amount of oxygen consumed in the oxycline layer is spent on oxidation of reduced forms of Fe, Mn, N. Deeper oxycline the amount of oxygen continues to decrease and at the depth of hydrogen sulfide (more than 0.005 mg H_2S/l) it still has a concentration of 0.1–0.3 ml/l [1].

The ecologically important question of the intensity and nature of deep Black Sea water movement continues to be the subject of a lengthy scientific discussion, the preliminary results of which can be summarized as instrumental data on the structure of currents in the entire water column accumulate. Multiday synchronous observations of currents in the 10–1500 m layer, performed at five autonomous buoy stations in the Main Black Sea Current (MBSc) zone, in the deep anticyclonic vortex and beyond these structures have yielded new interesting results [2]. First, currents at all depths are non-stationary, and among periodical measurements of the velocity vector vibrations with periods of about 17 and 4 h prevail. Secondly, vertical structures of currents differed by unidirectionality in the zone of the MBSc up to the layer 300–350 m, in a deep anticyclonic vortex—up to the horizon 1000 m. The vertical shift of the circular current velocity between the horizons 500 and 1000 m reached 5 cm/s, in the zone of MBSc between these horizons the shift of velocity was close to zero.

A series of quasi-synchronous surveys of the entire sea area with a grid mesh size of 20 miles of deepwater stations made it possible by calculating geostrophic currents and analyzing the distribution of properties of seawater to detect a fairly dense packing of cyclonic and anticyclonic dynamic structures (vortices and meanders), and the MBSc looks like an intense jet that meanders between vortices of opposite rotation. Below the 500 m horizon, water circulation consists of separate cycles, the most intense of which are associated with deep anticyclonic vortices [3]. In the case of the above described intensive counterflow measured at a depth of 1500 m, where the right (northern) part of the MBSc jet is simultaneously an element of the anticyclonic vortex in which a more powerful layer is involved in the circular motion. There is a hypothesis that a mixture of Lower Bosporus and Black Sea waters may spread across the Black Sea in several intermediate layers, but this hypothesis has not yet been confirmed by natural observations. Oceanological conditions in the northwestern part of the sea will be discussed in the next section of this paper.

7.2 Black Sea Hydrogen Sulfide: Geochemical Features of Behavior, Sources, Forms of Location, Dynamics and Trends of Variability

Hydrogen sulfide in the deep layers of the Black Sea waters was first discovered in 1890 by the Russian oceanographic expedition led by Academician I.I. Andrusov. Further studies have shown that this gas is present throughout the deep waters of the Black Sea, approaching the surface up to 80–100 m in the central areas of the sea and going down to 150–200 m in the zone of the continental slope and to the far shelf. This difference in the position of the upper boundary of the hydrogen sulfide zone is due to the specificity of water mass circulation in the sea, where there is upward movement of water in the center of the sea and downward movement of water at its periphery.

In the geological history of the Black Sea, the formation of hydrogen sulfide in its depths has always been associated with the penetration of saltier Mediterranean waters through the Bosporus Strait. At the same time, a significant volume of river runoff also enters the sea, resulting in a sharp density jump—pycnocline—between the distributed surface waters and the salty deep waters, making it difficult to mix the water masses vertically. As a rule, the upper boundary of a hydrogen sulphide zone begins at once under pycnocline, preventing inflow of oxygen from the top layers into this zone. During the climatic fluctuations of the ocean level, the connection of the Black Sea with the Mediterranean through the Bosporus Strait has been disrupted and renewed. The last time it recovered was 6,000 to 7,000 years ago. During this time, a water column containing hydrogen sulfide was formed in the Black Sea, in dynamic balance with the upper aerobic water column.

Marine Hydrophysical Institute of RAS constantly carries out researches of hydrogen sulphide zone of the Black Sea, both experimental and theoretical methods. By present time MHI RAS has the most representative database on distribution of hydrogen sulfide in Black Sea waters. This database covers the period starting from the 20 s. Total number of stations, at which measurements of hydrogen sulfide concentration distribution in water column were carried out, exceeds 4000 [4].

Generalization of materials of researches of a hydrogen sulfide zone of the Black Sea was carried out by known oceanologist Skopintsev [5]. He analyzed all available materials on this issue up to 1965, i.e., before the beginning of the development of the sea eutrophication process, which has now spread to almost all its water area. Even in those years B.A. Skopintsev suggested that if organic matter flow into the Black Sea increases (for example, due to increase of its biological productivity or greater inflow of low-proof organic compounds entering the sea with river waters), the chemical composition of the sea will change. These assumptions are now beginning to be partially justified. Not only in coastal waters, but also in open sea waters, an excess of organic matter has been found. There is no doubt that this situation will sooner or later affect the balance of hydrogen sulfide in the sea. To what extent this balance is determined by the influence of natural factors, and in what anthropogenic—an important fundamental problem of modern oceanology, a certain contribution to the solution of which is the work [6]. As a result of expeditionary researches of last decades the interseasonal and intraseasonal variability of hydrogen sulfide zone boundary has been defined. The closest to the surface (70–80 m) the upper boundary of a hydrogen sulfide zone is in the spring in area of a uniform cyclonic cycle in the center of the sea. In summer and autumn, in the presence of two stationary cyclonic cycles in their center, the upper boundary of the hydrogen sulfide zone is 95–110 m deep. At the periphery of the cyclones in all seasons of the year, a depth of 150–190 m is observed. Data on interannual variability of the hydrogen sulfide zone boundary strongly depend on duration of a time interval. Thus, changes in the position of this boundary over a fairly long period (about 60 years), its average depth has changed little according to estimates [4]. Experts note that against the background of registered interannual variations there is no constant unidirectional change of position of the hydrogen sulfide zone border [7]. The highest positions of the zone recorded during the study period, as a rule, were very short and local and were caused by active synoptic disturbances.

It is necessary to notice that the concept "the top border of a hydrogen sulphide zone" is rather conditional and is defined by set of difficult controllable factors. The top border is depth on which, according to the accepted technique in water samples presence of hydrogen sulfide with concentration about 0.1 ml/l is revealed. Position of the top border depends on speed of reaction of hydrogen sulfide oxidation, speed of delivery (thanks to vertical water exchange) of oxygen from the top and hydrogen sulfide from the bottom layers in an intermediate layer where oxidation occurs. Now on this account there is one more opinion [8] according to which in processes of hydrogen sulfide oxidation the considerable role is given to waters of the so-called "Bosporus mixture" at the expense of which the horizontal stream of oxygen providing oxidation about 40% of hydrogen sulfide, and sometimes more is formed.

Three main sources of hydrogen sulfide occurrence in reservoirs of hydrosphere of the Earth are known. Hydrogen sulfide is formed, firstly, at the expense of restoration of sulphates present in water at oxygen-free decomposition of organic substances. This process is carried out with the participation of sulfate-reducing bacteria, which use oxygen sulfates in the processes of its vital functions. Secondly, due to the rotting of organic compounds containing sulfur, and finally, thirdly, hydrogen sulfide can come from the depths of the Earth's crust through the clefts of the bottom and with hydrothermal waters. Thus how much appreciable concentration of hydrogen sulfide will be in this or that reservoir depends on speed of its oxidation. Hydrogen sulfide is oxidized by atmospheric oxygen soluble in sea water and oxygen formed as a result of photosynthesis of algae. Oxidation can also be carried out microbiologically through the vital activity of sulfur dioxide microorganisms with the help of seawater oxygen, and in the absence of the latter with the help of oxygen nitrates. Thus, the balance of arrival and consumption of hydrogen sulfide and determines its appreciable presence in this or that water body. The most extensive hydrogen sulfide infestation is observed in the Black Sea basin. Thus the source of hydrogen sulfide connected with rotting of sulfur-containing organic substances, by modern estimations is insignificant and makes 2–5% from available hydrogen sulfide in the Black Sea.

As possible sources of hydrogen sulphide in the Black Sea are considered:

- anaerobic destruction of organic matter by sulfate reducing bacteria;
- inflow of hydrogen sulfide from the bowels of the Earth through fissures of the sea floor in the form of gas and with hydrothermal solutions.

At present it is confirmed that microbiological sulfate production is the main source of hydrogen sulfide in the Black Sea. Thus the main reasons of existence of a hydrogen sulfide zone in the sea are considered to be the density stratification hindering a vertical exchange and the big biogenic runoff from coast in calculation on unit of the sea area. Both factors are under strong anthropogenic influence: regulation of river runoff and strengthening of biogenic runoff. Since most of the organic matter in the Black Sea is formed on the shelf, the ecosystem of the latter largely determines the state of the sea's hydrogen sulfide zone [7]. The energy exchange of sulfatered microorganisms is based on anaerobic oxidation of low molecular weight organic substances to CO_2 due to sulfate reduction. Study of the mechanism of bacterial sulfate reduction showed that the reduction of sulfates as well as the oxidation of hydrogen sulfide occurs in several stages and is accompanied by the formation of various compounds associated with the cycle of sulfur in the sea [6] (Fig. 7.1).

Fundamental work on sulfate reduction in the Black Sea was carried out by Sorokin [9], who for the first time experimentally determined the activity of bacteria, their number and rate of formation in various layers of the water column. They also proved that about 50% of hydrogen sulfide is formed in the surface layer of

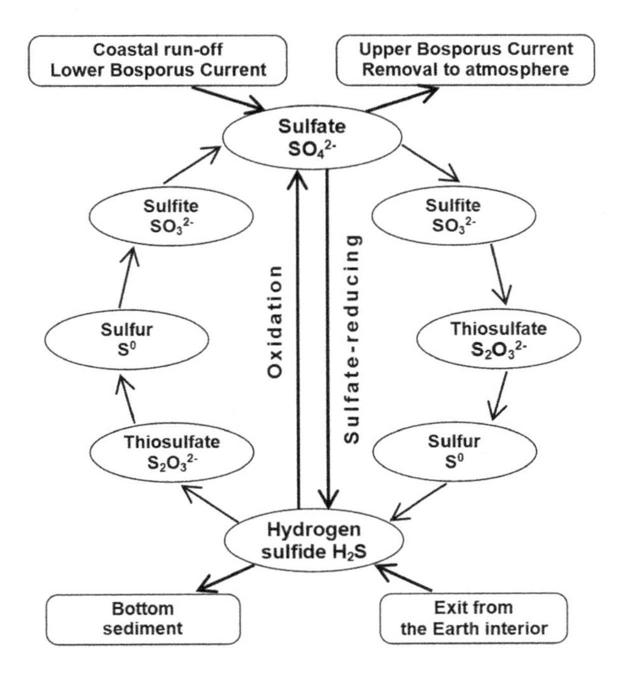

Fig. 7.1 Scheme of the processes of oxidation and recovery of sulfur in the Black Sea

precipitation of the hydrogen sulfide zone. The results of recent studies have shown that the process of bacterial sulfate reduction occurs in the entire water column of the hydrogen sulfide zone, as a result, integrated products for the entire sea according to Bezborodov and Eremeev [10] was 6×10^7 tons/year. The focal character of sulfate reduction was established due to zoning of easily assimilable forms of organic matter entering the sea.

Experimental work has established that in the contact zone of aerobic and anaerobic waters in the Black Sea can simultaneously take place the processes of bacterial reduction of sulfates, producing hydrogen sulfide and the processes of oxidation of formed hydrogen sulfide (Fig. 7.1) [6, 11]. Currently, there are also convincing evidence that the oxidation of hydrogen sulfide is carried out both by chemical and microbiological means.

Considering a modern condition of a hydrogen sulfide zone of the Black Sea it is necessary to stop on influence on its dynamics and evolution of anthropogenic pollution. The increase of biogenic elements as a result of industrial, domestic and agricultural pollution leads to an increase in production of dead organic matter, stimulating the process of sulfate reduction and hydrogen sulfide replenishment. At the same time in the aerobic zone oxygen is spent on decomposition of additional amounts of organic matter, which reduces the possibility of rapid oxidation of hydrogen sulfide in the case of its local inputs from the underlying sea layers. Since most of the organic matter in the Black Sea is formed on the shelf, processes in its ecosystem largely determine the state of the hydrogen sulfide zone of the deep sea. According to rough estimates, due to anthropogenic pollution in the Black Sea already today may arise additional amounts of hydrogen sulfide, comparable to the formed naturally.

Thus, the Black Sea can be considered as an ecosystem consisting of four zones: aerobic, microaerophilic, anaerobic and silt zones. The hydrogen sulphide zone is not dead water, but a bacterial ecosystem well balanced in its functions with the sea's aerobic ecosystems. Its bacterial population provides the carbon and nutrient cycle as good as the deep sea ecosystems without hydrogen sulfide.

Proceeding from positions of consideration of the Black Sea hydrogen sulfide zone as a difficult bacterial ecosystem, in work [12] the mathematical model of an ecosystem of a hydrogen sulfide zone which purpose—forecasting of possible consequences of evolution of a hydrogen sulfide zone in the conditions of increasing anthropogenic loadings is presented.

7.3 Features of Vertical Distribution of Nitrogen and Phosphorus—The Main Biogenic Elements that Provide Bioproductivity of Marine Ecosystems (Modern Concepts)

Vertical phosphorus distribution

The presence of two zones with different gas regimes (oxygen and hydrogen sulphide) in the Black Sea determines the specificity of vertical distribution of the main biogenic elements—nitrogen and phosphorus in the sea in comparison with aerated waters of the World Ocean.

The average phosphorus content in Black Sea waters is 3–4 times higher than in the oceans. Its minimum concentration is observed in the upper 50-m layer, with the depth increasing, on the horizons of 500–2000 m the concentration of phosphorus is almost 20 times higher than in the upper layer.

In the surface layer, the distribution of phosphate coincides with the flow pattern of the Black Sea water. Maximum phosphate content is observed in the centers of cyclonic cycles, due to the rise of deep waters enriched with phosphates. In anticyclonic structures, the waters are depleted of phosphate due to their sinking. For this reason, minimum phosphate concentrations are present in waters at the periphery of the sea. The phosphate content is subject to considerable inter-seasonal variability and depends on the meteorological situation. During cold storms, vertical water mixing increases and the upper phosphate layer is enriched with phosphate.

Intensification of hydrochemical research in the Black Sea in the 80s and 90s allowed to make some corrections to the idea of vertical distribution of phosphorus in the sea, especially in the contact zone of aerobic and anaerobic waters, in comparison with the data of Skopintsev [5], according to which the curve of vertical phosphate distribution was represented as a continuous increase of phosphorus concentration from the lower boundary of the upper homogeneous layer up to depths of 1000–2000 m.

A qualitatively new level of measurements of phosphates in the 80–90s allowed the authors of the works [13] to find a sharp minimum of phosphates on the upper boundary of the layer of coexistence of oxygen and hydrogen sulfide (Fig. 7.2), the concentration of which sometimes falls to analytical zero.

The existence of a minimum of phosphates on the upper boundary of the contact layer of aerobic and anaerobic waters is associated either with the processes of chemosynthesis on reduced forms of sulfur [14], which is confirmed by the presence in this layer of maximum concentrations of organic phosphorus (up to 0.6 μg-atm/l), or by physical and chemical sorption processes on freshly formed hydrooxides Fe^{+3} and Mn^{+4}, as evidenced by the coincidence of minimum phosphates with maximum suspended Fe^{+3} and Mn^{+4} and a sharp increase in suspended phosphorus on almost the same horizons.

Thus, the detected anomalies in the vertical distribution of phosphates belong to a rather narrow layer of water in the contact zone of aerobic and anaerobic water masses and are associated with the peculiarities of production and destruction processes of organic matter occurring in this zone, in which a significant place is occupied by bacterial chemosynthesis.

Expeditionary Researches of southern science center (SSC) RAS in 2013 [15] showed that the regularities of vertical distribution of mineral and organic forms of nitrogen and phosphorus indicate that in the deep anaerobic zone of the Black Sea there is a transformation of organic matter associated with its deep destruction and release of mineral forms of biogenic elements. It was found that at the depths of 100 m and below, the Black Sea waters contain a significant stock of mineral

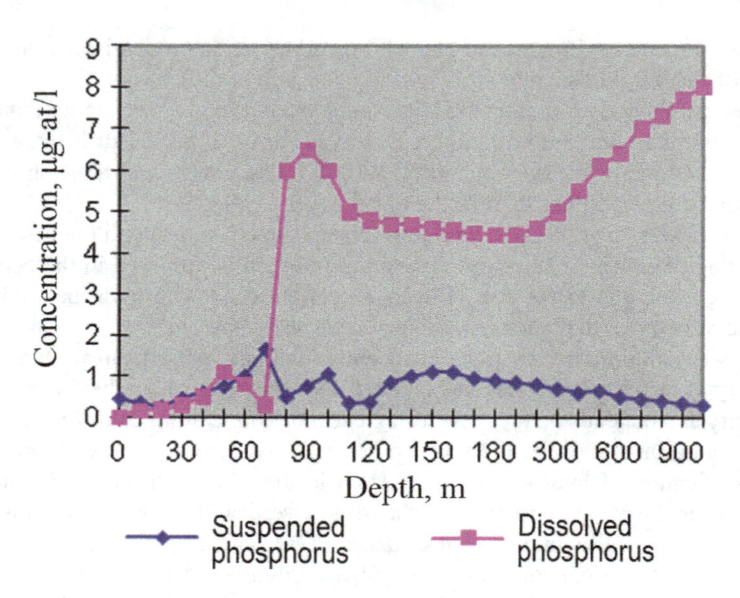

Fig. 7.2 Vertical distribution of dissolved and suspended phosphorus

phosphorus, the transfer of which into the photonic zone is possible during upwelling. Low concentrations of phosphorus (0–3 µgP/L) have been observed in the surface sea layer. At a depth of 100 m phosphorus concentration is significantly increased and further increases with depth, which is probably due to accumulation of settling organic matter and subsequent bacterial mineralization in the anaerobic zone [16].

Classical notions of Horn [17] about the peculiarities of the phosphorus cycle in marine ecosystems and the role in this cycle of exchange at the water-bottom boundary, in the conditions of the deep hydrogen sulfide zone ecosystem acquire somewhat different meaning. In conditions of the Black Sea at consideration of a phosphorus cycle the hydrogen sulfide zone actually carries out a role of soil as the phosphorus adsorbed in overlying layers of water on hydroxide of trivalent iron, at hit of the last in a hydrogen sulfide zone, there is a transition of iron in a two valence condition and as a result—desorption of phosphorus and a maximum on a curve of its vertical distribution.

Vertical distribution of nitrogen mineral forms and their flows in the Black Sea Among mineral forms of nitrogen in the Black Sea ammonium prevails In waters of the Black Sea distribution of forms of nitrogen on depth has the important feature—considerable growth of a share of mineral nitrogen (to 47%) at the expense of increase in concentration of ammonium with approach to a hydrogen sulphide zone.

The lowest average content of ammonia nitrogen is observed in the surface layer, where it is formed as a result of the decomposition of organic compounds, but after the use of this form of phytoplankton, seasonal fluctuations in its content in this layer are observed. At 200 m the concentration of ammonia nitrogen is almost an order of magnitude higher, at 2000 m it reaches 1300–1400 mgN/m^3.

The presence of large and increasing concentrations of ammonia nitrogen in the Black Sea is mainly due to the processes of decay of sedimentary organisms in the water column and at the bottom. This process, of course, takes place at the bottom of the seas and oceans, but in them ammonification (the first stage of mineralization of nitrogen-containing organic matter) is accompanied by the subsequent stage—nitrification, resulting in nitrates (through a brief stage—the formation of nitrites).

Ammonia nitrogen can also be formed as a result of the life activities of denitrifier bacteria that restore nitrites and nitrates. The maximum content of ammonia nitrogen, as well as hydrogen sulphide, is observed at the bottom and this may indicate that the bulk of ammonia nitrogen, formed at the bottom, is distributed in the water due to vertical movement of water. At the top border of the hydrogen sulfide zone is its oxidation to nitrates, and in a layer of photosynthesis this form of mineral nitrogen can be used by phytoplankton. Characteristically, that and the top site of a curve of distribution of ammonia nitrogen (Fig. 7.3a) deviates from an axis of ordinate to the right as it takes place and for hydrogen sulfide. This confirms that the formation of ammonia nitrogen also takes place in the upper layer of water. Coincidence of a course of curves of hydrogen sulphide and ammonia nitrogen gives the basis to believe that distribution of ammonia nitrogen in a hydrogen sulphide zone (where it is not oxidised to nitrates and is not consumed by phytoplankton), as well as hydrogen sulphide, is caused by diffusion.

Nitrate nitrogen. In the upper water layer, especially during the vegetation period, nitrates are usually absent. Seasonal variations of nitrates can be presented as follows: in warm seasons they are absent in the layer from 0 to 50–75 m, below they are found up to the horizon of 175 m. In winter, the nitrate content in surface waters can reach a maximum.

Deeper than the photo layer, nitrate concentration increases as a result of nitrification, but the increase in nitrate concentration is limited to a depth of 150–175 m. Below 200 m nitrate content sharply decreases due to denitrification and thiodenitrification processes occurring in the hydrogen sulfide zone.

The vertical distribution of nitrogen mineral forms shows that microbiological processes play a major role in the transformation of these forms, which can be built into the next chain:

Upper water layer	–	*Nitrification* $NH_4^+ \rightarrow NO_2^- \rightarrow NO_3^-$ Oxidation and its absorption by phytoplankton
Upper water layer above the H_2S area	–	*Denitrification* $5\,CH_3OH + 6\,NO_3^- = 3\,N_2 + 5\,CO_2 + 7\,H_2O + 6\,OH^-$, Oxidation of organic matter by nitrates to form molecular nitrogen N_2

(continued)

(continued)

Layer of contact of aerobic and anaerobic waters, lower boundary	–	*Tiodenitrification* $S_2O_3^{2-} + 2NO_3^- + 2H^+ \rightarrow 2SO_4^{2-} + N_2 + H_2O$ $5 H_2S + 2 NO_3^- + 2 H^+ \rightarrow 5 S^O + N_2 + 6 H_2O$ oxidation of H_2S and $S_2O_3^{2-}$ by nitrates—molecular nitrogen is formed
Microbiological process of N_2 fixation	–	Nitrogen fixing microorganisms can be aerobic, anaerobic and microaerophilic. The fixation process is optimal at 0.14 mgO_2/l
Under anaerobic conditions, methanobrazuyuschie and sulfatereducing bacteria can fix nitrogen		
Hydrogen sulfide zone		*Ammonification*—decomposition of organic matter with the formation of ammonium

It is known that the cycles of the two listed biogenic elements of nitrogen and phosphorus are not closed, the difference between them can be indicated as follows: for nitrogen the more important in the cycle is the exchange of atmosphere—sea, and for phosphorus—on the border of water—bottom. If we consider these differences in the ecosystem of the deep hydrogen sulfide zone, then for nitrogen denitrification processes will take place mainly within the suboxygen zone at significant depths and it would seem that this can lead to a significant oversaturation of water with molecular nitrogen N_2, but correct data confirming this process is not yet available. Obviously, this is due to the fact that at about the same depths can simultaneously take place the processes of sulfate reduction and microbiological oxidation of reduced compounds such as S^O, $S_2O_3^{2-}$, H_2S, CH_4, as is known in the Black Sea. Sulfatereducing bacteria and a number of populations of methane and sulfur oxidizing microorganisms are active nitrogen-fixing agents, therefore, the accumulation of nitrogen in the cells of these microorganisms may occur, and their subsequent extinction and decomposition may be another reason for increasing the concentration of ammonium in the hydrogen sulfide sea zone.

Assessment of flows and redistribution of mineral forms of nitrogen in the Black Sea water column, including its aerobic and anaerobic zones, is very relevant due to the increasing processes of sea eutrophication in the last three decades due to anthropogenic supply of biogenic elements, including nitrogen.

In this paper [18] on the basis of averaging the data of long-term observations on the vertical distribution of nitrates and phosphates, as well as calculations of rates of excretion of biogenic elements by plankton organisms, estimates of the average monthly values of upward and downward flows of inorganic nitrogen and phosphorus compounds in the photosynthesis zone of the deep Black Sea region were obtained. The upward flow calculations assumed that vertical transport rates were controlled by density gradients at the upper limit of the main pycnocline throughout the year. It was found that nutrient inputs to the photosynthesis zone due to physical processes gradually increased throughout the year from the minimum values in July and August (0.1–0.3 mg at N m^{-2} day^{-1} and 0.02–0.04 mg at P m^{-2} day^{-1}) to the maximum values in February and March (1.2–1.8 mg at N m^{-2} day^{-1} and 0.2–0.3 mg at P m^{-2} day^{-1}). Seasonal dynamics of the regeneration flow has an

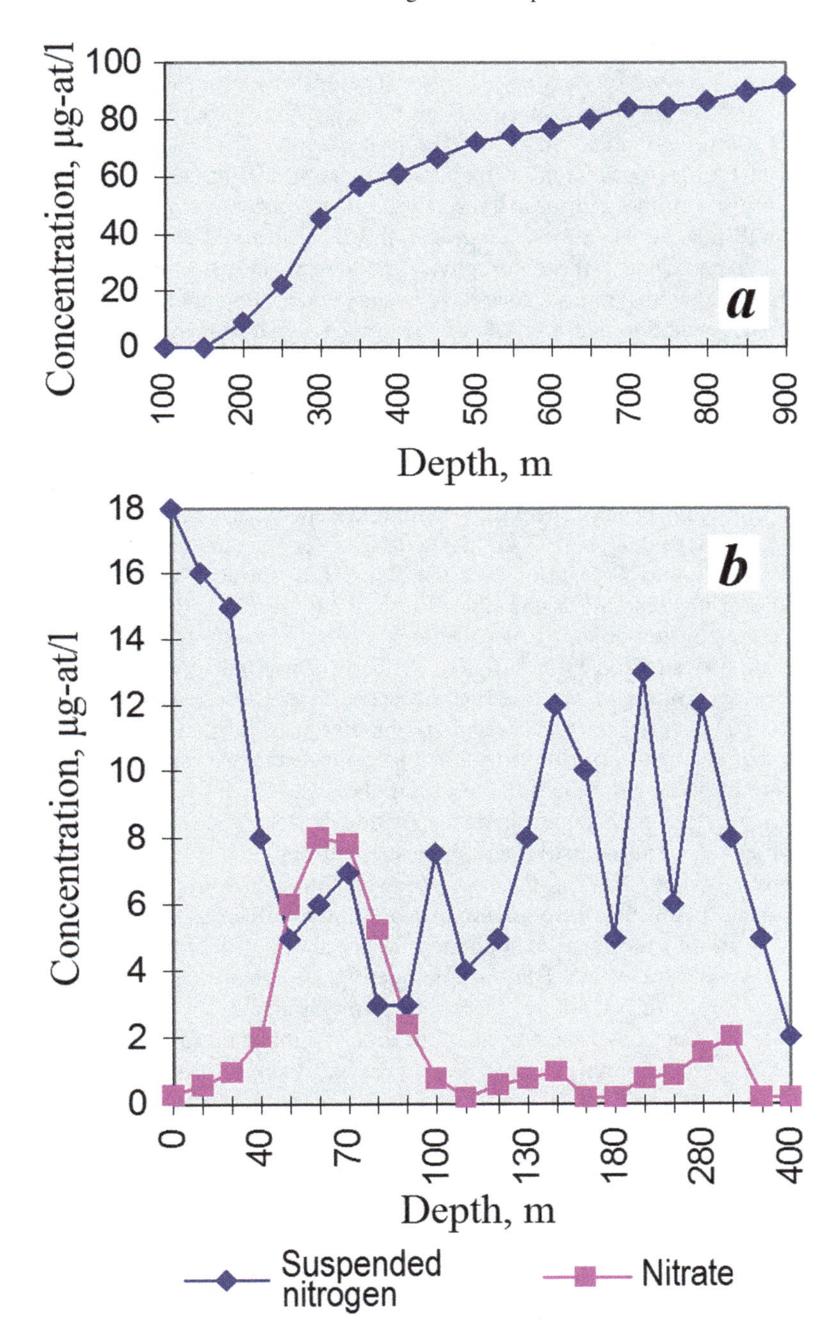

Fig. 7.3 Vertical distribution of inorganic nitrogen in the Black Sea: ammonium (**a**), nitrate and suspended (**b**) according to data [10]

oppositely directed annual course. In summer, regeneration rates increase by approximately 5 times compared to winter. The upstream contribution to the total nutrient inputs to the photosynthesis zone (F-ratio) varies throughout the year from 5 to 50% (nitrogen) and from 10 to 70% (phosphorus). It is shown that seasonal changes of F-value correspond to the annual dynamics of the share of "new" and export products in the total primary synthesis of the substance in the pelagic sea. The natural links between the average monthly F values and the concentration of nitrates and phosphates, the chlorophyll "a" content and the biomass of phytoplankton in the photosynthesis zone have been revealed. Potentially possible values of "new", regenerative and total phytoplankton production have been determined. Correspondence between average monthly and annual estimates of primary production calculated on the basis of biogenic element flows and averaged data of in situ measurements is shown.

The paper [19] presents a graph of the dependence between the amount of precipitating soluble forms of nitrogen and phosphorus per hectare per year in some areas of the world ocean. There is a symbiotic dependence between these forms according to Savenko [19]. With dusty and watery sediments, total phosphorus $\sim 22 \times 10^3$ tons P/year falls on the Black Sea surface. The soluble forms are $\sim 50\%$ of this amount of phosphorus deposited. Using the dependencies given in [19] the estimates obtained in [6] that 11×10^3 tons P/a of soluble forms of phosphorus deposited on the Black Sea surface correspond to 150^{103} tons R/a of soluble forms of nitrogen deposited on the sea surface. Close to the obtained value of 135×10^3 tons N/year is obtained by another way [20], if the average concentration of soluble forms of nitrogen in a liter of coastal rainwater is multiplied by the amount of precipitation falling out during the year. Thus, for the Black Sea, the supply of nitrogen with atmospheric deposition is 26.5%, the supply with river discharge is 44%, and industrial and domestic sources 29.5%.

According to data [21–23], the main forms of nitrogen from precipitation to the surface of the Black Sea were ammonium and nitrates. The average annual background flow of nitrogen compounds was about 0.65 tons/(km^2 year) or 46.5×103 mol/(km^2 year). During precipitation, inorganic nitrogen is gradually spreading to a depth. Therefore, the effect of precipitation on inorganic nitrogen content in the sea can be expected to be more significant for the upper layer. However, the relative importance of this effect will increase from coastal areas to the open sea, as the average inorganic nitrogen content of surface water decreases from coastal to open sea. At total volume of atmospheric precipitation of 230 km^3, mineral forms of nitrogen in them make 136,850 tons, from them on ammoniacal nitrogen it is necessary 87,500 tons or 64%. The domination in atmospheric depositions of the ammoniacal form of nitrogen which, as it is known, is preferable in a food of phytoplankton, allows to assume that for the Black Sea such natural-climatic factor as atmospheric precipitation can to some extent influence level of eutrophication of an ecosystem of an open part of the sea.

The main features of inorganic nitrogen distribution in the surface layer of the sea are that their content in the north-western part, which is affected by coastal runoff, is high in both cold and warm seasons due to the inflow from the rivers. In open sea areas the concentration of inorganic nitrogen is much lower than in coastal and nearshore areas. At the same time, during the cold period of the year the content of inorganic nitrogen compounds in this part of the sea exceeds the content during the warm period due to surface waters with increased content of inorganic nitrogen from the Azov Sea, more intensive, in comparison with the summer period, physical exchange with waters of the main pycnocline, which have increased content of nitrates and ammonium.

Among the mineral forms of nitrogen in the Black Sea, ammonia (~99%) prevails, so it is from this form that we will consider its flows in all layers of sea water. The lowest ammonia nitrogen content is observed in the surface layer, where it is formed due to the decomposition of organic compounds and used by phytoplankton, causing seasonal variability of its content in this layer of water. According to the data [13] on the sea surface (0 m) the NH_4 + content varies from 1.4 to 80 μg N/l, with an average NH_4 + content of 5.9 μg N/l in winter and 2.5 μg N/l in summer. At a depth of 200 m, the ammonium nitrogen content increases by an order of magnitude, reaching 1300–1400 μg N/l at a depth of 2000 m, and 1400 μg N/l at the bottom [13].

The flows of ammonia nitrogen in the Black Sea are presented in the form of a scheme in the paper [6], where the flows on the border of I—the surface water layer; II—the deep water layer; III—the bottom water layer are considered. The scheme takes into account all the articles of the parish: river runoff; atmospheric precipitation; Lower Bosporus flow; and articles of discharge: evaporation; Upper Bosporus flow; and exchange between layers I, II, and III. The scheme provides a combination of three figures: total water volume; NH_4^+ concentration; and total NH_4^+ flow. It follows from the scheme that the NH_4^+ flux from external sources is 143,525 tons of ammonia nitrogen per year. This value is very small compared to the ammonium nitrogen content of the Black Sea water column, which for a deep water mass of 419,000 km^3 is 305 × 10^6 tons NH_4/a, and taking into account the ammonium nitrogen contained in the bottom water layer, 83 × 10^6 tons NH_4/a will be 388 × 10^6 tons of ammonium nitrogen.

The scheme of nitrate flows in all layers of the Black Sea water is presented in the paper [6]. The scheme of ammonia flows is considered in a similar way: I—surface water layer, II—deep water layer, III—bottom water layer. It is shown that the amount of nitrates entering the Black Sea from external sources (river runoff, atmospheric precipitation and the Lower Bosporus flow) is 380,000 tons or 0.38 × 10^6. This amount is almost commensurate with the nitrate content of the surface water layer (0.3 × 10^6 tons) and accounts for 60% of the nitrates produced in the deep sea water layer. The information obtained on flows of inorganic forms of nitrogen to various layers of Black Sea water has shown that of all the forms of nitrogen entering the sea from external sources, flows of nitrates pose the greatest threat to the ecosystem. Already now, given the current ecological state of the sea, these flows are commensurate with the amount of nitrate nitrogen produced in the surface water layer and account for 60% of the nitrates contained in the deep sea layers. If the total amount of all mineral forms of nitrogen entering the sea from external sources is estimated, it is 528,350 tones of nitrogen. The atmospheric precipitation of these forms is 136,850 tons or 26.5% of the total input.

Study of the fine hydrochemical structure of the layer of interaction of aerobic and anaerobic waters (including biogenic elements of nitrogen and phosphorus) in the works [6, 10] made it possible to establish that this layer is a chemical and microbiological reactor, where the rate of geochemical and microbiological processes is determined by the rate of delivery of reacting components both from the overlying photo layer of water and from the deep hydrogen sulfide zone, as well as by the peculiarities of production and destruction of organic matter through microorganisms of biogeochemical cycles of sulfur, nitrogen, manganese, iron and methane.

The differences in flows of inorganic forms of nitrogen such as ammonium and nitrate nitrogen are shown. Flows of the ammoniac form of nitrogen into the surface layer of sea waters are determined not by the volume of ammonium coming from external sources, but by its flows from the deep and near-bottom layers of sea waters, and streams of nitrates—by their supplies from external sources.

Insufficient study of fundamental processes occurring in Black Sea ecosystems not only prevents a reasonable forecast of their evolution, but also makes monitoring of these ecosystems ineffective.

7.4 Conclusions

A brief review of modern views on the features of circulation processes in the Black Sea and their impact on the formation of the hydrochemical regime of the Black Sea, the causes of hydrogen sulfide zone formation in the sea is given.

The hydrogen sulfide zone of the Black Sea as a bacterial ecosystem, well balanced in its functions with sea aerobic ecosystems, the main mechanisms of which are determined by chemical and microbiological processes of transformation of inorganic forms of sulfur and nitrogen. Its bacterial population provides the cycle of biogenic elements and organic carbon in the sea ecosystem as a whole, not worse than the deep sea ecosystems without hydrogen sulfide.

By means of averaging the data of long-term observations on the vertical distribution of nitrates and phosphates, as well as calculations of rates of excretion of biogenic elements by plankton organisms the estimates of average monthly values of upward and regenerative flows of inorganic nitrogen and phosphorus compounds in the zone of photosynthesis of the deep Black Sea were obtained.

The flows of mineral forms of nitrogen (NH_4^+, NO_3^-, NO_2^-) in the deep part of the sea have been estimated taking into account their inflow from external sources and the processes of exchange between layers (surface, deep and near-bottom) of water in the sea.

Relevance of obtaining fundamental knowledge about the mechanisms of functioning of marine ecosystems, to what extent these mechanisms are determined by anthropogenic, and in what natural and climatic factors are also related to the need for scientific justification of environmental activities in the region.

References

1. Ivanov VA, Pokazeev KV, Sovga EE (2006) World ocean pollution. Moscow State University. 163 p
2. Oquz T, Latun VS, Latif MA, Vladimirov VV, Sur HI, Arkov AA, Ozsoy E, Kotovshcikov BB, Eremeev VN, Unluata U (1993) Cirrulation in the surface and intermediate, laqers of the Blak Sea. Deep-Sea Res 40(8):1597–1612
3. Oguz T, Aubrey DG, Latun VS et al (1994) Mesoscale circulation and thermohaline structure of the Black Sea observed during Hydro Black 91. Deep Sea Res I 41(4):603–628
4. Eremeev VN, Suvorov AM, Halliulin AH, Godin EA (1996) About correspondence of position of the upper boundary of hydrogen sulfide zone of the certain isopic surface in the Black Sea by perennial observations. Okeanologiya 36(2):235–240
5. Skopintsev BA (1975) Formation of the modern chemical composition of the Black Sea waters. Leninhrad. Gidrometeoizdat, 336 p
6. Sovga EE (2002) Features of mechanisms of functioning of the Black Sea ecosystems of the shelf and pelagiali. The abstract of the dissertation for a scientific degree of doctor of geographical sciences. MGI NAS of Ukraine, Sevastopol, p 37
7. Belyaev VI, Sovga EE (1991) Hydrogen sulfide will not explode in the Black Sea. Newsletter of the USSR Academy of Sciences, № 10, pp 47–57
8. Eremeev VN, Ivanov LI, Konovalov SK, Samodurov AS (2001) Role of oxygen, sulfides, nitrates and ammonium flows in the formation of hydrochemical structure of the main pycnocline and anaerobic zone of the Black Sea. Marine Hydrophysis l:64–82
9. Sorokin YI (1982) Black Sea. Science, Moscow, p 216
10. Bezborodov AA, Yeremeev VN (1993) Black Sea. Interaction zone of Aerobic and Anaerobic waters. MGI NAS of Ukraine, Sevastopol, 298 p
11. Sovga EE, Eremeeva LV, Solovyova LV (1989) Geochemical features of transformation of inorganic forms of sulfur in the Black Sea water column. Geochemistry 11:1648–1655
12. Belyaev VI, Sovga EE, Lyubartseva SP (1997) Modelling the hydrogen sulphide zone of the Black Sea. Ecol Model 96:51–59
13. Sapozhnikov VV, Gusarova AI (1990) Anomaly of vertical phosphate distribution in the Black Sea. Oceanographic aspects of protection of seas and oceans from chemical pollution. Gidrometeoizdat, pp 147–149
14. Vinogradov ME, Sapozhnikov VV, Shushkina EA (1992) Black Sea ecosystem. Nauka, Moscow, p 112
15. Matishov GG, Stepanyan OV, Grigorenko KS, Kharkovsky VM, Povazhny VV, Sawyer VG (2013) Peculiarities of hydrologic and hydrochemical regime of the Azov and Black Seas in 2013. Vestnik of the south sciences center 11:36–44
16. Sorokin Yu I (2002) Black Sea ecology and oceanography. Backhaus Publishers, Amsterdam, 875 p
17. Horn R (1972) Marine chemistry. Ezd. World, 398 p
18. Krivenko OV, Parkhomenko AB (2014) Upward and regenerative flows of inorganic nitrogen and phosphorus compounds in the deep Black Sea region. J Gen Biol 75(5):394–408
19. Savenko VS (1996) Phosphorus in atmospheric precipitation. Water Res 23(2):189–199

20. Rozhdestvenskiy AV (1998) Geochemistry and hydrochemistry of the Danube drain and the atmospheric precipitation in connection with the modern sediments of the Black Sea (in Russian) (vol 2). Institute for Oceanology, Varna, pp 20–26
21. Varenik A, Konovalov S, Stanichny S (2015) Quantifying importance and scaling effects of atmospheric deposition of inorganic fixed nitrogen for the eutrophic Black Sea. Biogeosciences 12:6479–6491
22. The Black Sea (1997) Transboundary diagnostic analysis. The global environment facility Black Sea environmental programme. Programme Cordination Unit, 180 p
23. Shoemakers VV (1990) Ammoniacal nitrogen in the Black Sea (in Russian). Oceanology 30 (1):53–58

Chapter 8
Biotic Characteristics (Flora and Fauna) of the Black Sea

8.1 Introduction

The Black Sea is a unique sea and the distribution system of animals and plants in it is very unusual. Only 13% of the total volume of water is inhabited by living organisms and then at a depth of no more than 200 m. Below this mark the water is saturated with hydrogen sulfide and only some species of anaerobic bacteria, which do not need oxygen, survive there. No other sea in the world has such a division into oxygen and hydrogen sulphide zones. Relatively small living space is the main reason why the flora and fauna of the Black Sea is not very diverse.

In total, there are more than 660 species of plants and 2500 species of animals in the Black Sea. The resources of the Black Sea, namely Hamsa, horse mackerel, sprat, mullet, flounder, mackerel, etc. are of industrial importance. Algae (phyllophora, cystosir, zostera, etc.) and invertebrates (mussels, shrimps, oysters). Every year the sea gives up to 300 thousand tons of biological resources. The plant world is represented by various algae and phytoplankton. There are certain species whose diet consists exclusively of organics. But the overwhelming majority "get their food" by the traditional method—photosynthesis. The animal world consists of 2.5 thousand species (for comparison, there are about 9 thousand species in the Mediterranean). Most of the Black Sea invertebrates are 1140 species. There are 500 single celled and crustacean species each, 200 shellfish and 160 vertebrates (fish and mammals) (https://vsya-planeta.ru/flora-i-fauna-chernogo-moria/).

8.2 The Flora of the Black Sea

Peridinium algae or dinoflagellates—single celled algae are part of phytoplankton, the presence of which in water can only be clearly seen during the "blooming" period. The water acquires a distinctly green tint. It is especially noticeable in bays

K. Pokazeev et al., *Pollution in the Black Sea*, Springer Oceanography,
https://doi.org/10.1007/978-3-030-61895-7_8

and estuaries. A special kind of peridinium algae—noctiluca or nochesworms—is an algae with phosphorescent powers. Noctiluca or nocttiliuca is a special type of peridinium algae. It lacks chlorophyll, so it feeds on ready-made organic substances. The algae has received its name—noctiluca—for its phosphorescent powers. Thanks to it in August you can observe the phenomenon of stunning beauty—the water begins to glow bluish shimmering and very mysterious light. Kladophora algae is represented by two subspecies—white and stray, preferring shallow waters. The first one covers wet stones with emerald velvet. The second one looks like a tangled coil of thin green threads. The stray one was named for the fact that it floats freely in water and does not try to get fixed in one place.

String algae also include transparent ceramium and bearded red or burgundy cystosyre. Transparent ceramium also refers to filament algae, but is red or burgundy. It grows in shallow water among stones. A large number of transverse lime rings support the shape of the plant. This gives the ceramium a funny, almost ornamental look: a bush of burgundy striped woolen threads with a loop at the end.

Cystosyrene Bearded Algae species, which forms whole thickets on hard soils. Sometimes its height is comparable to that of a man. Among its branches there are many small animals (fish, mussels, worms, etc.). Epiphytes of parasitic algae like to settle on the cystosir itself.

A zooster or a jackpot is a grass that has adapted to life under water. Grows on the sand, away from the shore. Interesting fact about Black Sea plants! Many algae are not only an important part of the ecosystem, but are also widely used by humans for their needs. For example, paint thickener (sodium alginate) is extracted from cystosyre, phyllophora is used in confectionery industry and pharmacology, dried zoster is stuffed with soft furniture (https://vsya-planeta.ru/flora-i-fauna-chernogo-moria/).

8.3 Black Sea Fauna

The Black Sea is home to a wide variety of animals—from unicellular to mammals. And although the diversity of fauna is much more scarce than in other seas, there are unique and very interesting species.

Two species of jellyfish live in the Black Sea—Aurelia and Cornerot. These animals are not hunted by anyone, because they are 98% composed of water and can not be food for a predator. But that doesn't mean the jellyfishes are useless. They're like filters, letting water through, purifying it from plankton. They are good weather forecasters—12 to 15 h before the storm starts, they leave the shore. Aurelia is shallower and absolutely safe for humans; it plays an important role in Black Sea plankton. Its population has increased dramatically in the 70s, 10–100 times more than in the 60s. Its population has eaten 34–67% of the total annual reserves of Black Sea plankton.

Scallop Double-winged clam, on the sea bottom at a considerable distance from the shore (up to 20 km). On land, the surf often throws away empty casements that look like a fan. Scallops "invented" a very original way of moving: opening and closing the sashes they, pushing out the water, jump. When the clam does not threaten anything, he releases sensitive tentacles that form a picturesque fringe along the edge of the shell. As the danger approaches, the tentacles hide and the scallop "flees".

Rapana For Rapana, the Black Sea is not its native land. This predatory mollusk "penetrated" into this pond through the Straits of Dardanelles and Bosporus and was first registered in 1947. In his new home he did not have his main enemies—the stars of the sea. The population therefore rapidly increased, causing considerable damage to the oysters and scallops that rapana feeds on. The mollusk is edible and tastes like sturgeon.

The mussels are another kind of edible clams. These bivalve mussels live by attaching themselves to stones, rocks and even pier supports. They scratch the water, extracting oxygen and useful organic matter from it. Black Sea pearls can form in them. Fried mussel meat is very tasty and very nutritious.

In the Black Sea there are 19 species of "beach attendants"—crabs, most of which are listed in the Red Book. The most common species that inhabits the coastal parts of the sea is the marbled crab. It can reach the land, but at the first danger it is immediately hidden in the water. Violet crabs are different from their marble counterparts by their sluggish and sluggish nature. That's why in case of a threat they freeze on the spot, trying to remain invisible to the enemy.

Stone crabs are deep-sea inhabitants, but at night they face the coastal rocks. This species is considered extinct as it is caught and eaten. From other congeners of the crab-rocks differ in large sizes, up to 7 cm. A close relative of a stone crab is a hairy crab. It differs from its congeners red-brown shell, which is covered with yellowish bristles and smaller sizes (about 3 cm). Young crab of this crab most often has a bright white color. It inhabits both in the coastline zone and at depths up to 35 m.

Grass crabs are also superior to other brethren, so they often come to the table for seafood lovers. Their shells have a trapezoidal shape, and to protect them from enemies, they use short but very strong claws. Brown crab is the largest and very rare species, which is on the verge of extinction. Its size can reach 20 cm.

In addition to large and medium-sized crabs, the Black Sea is home to many small species that look very similar to spiders. Among them is a **long-legged crab**, interesting because its color depends directly on the place of residence. All his life, this species hides in algae. Another inconspicuous inhabitant of the Black Sea waters is the crab phalanx https://moreprodukt.info/kraby/chernogo-morya.

Crustaceans are a very important part of the Black Sea fauna. They are considered a kind of marine corpsmen, as they clean the bottom and water by eating dead organisms.

8.3.1 Black Sea Fish

The Black Sea is home to 184 species and subspecies of fish, 144 of which are exclusively marine, 24 are passing or partially passing, and 16 are freshwater. In recent years, ichthyocenosis of the Black Sea has been supplemented by the Far Eastern mullet Pelengas Mugil so-iuy Basilewsky, successfully acclimatized in the Azov-Black Sea basin [1, 2].

Black Sea marine fish species are divided into 4 groups: permanently inhabiting (Black Sea race Khamsa, Black Sea horse mackerel, Black Sea sprat, kalkan); wintering in the Black Sea, but spawning and feeding in the Azov Sea (Azov race Khamsa, Kerch race herring); Wintering and spawning in the Black Sea, but fattening in the Azov Sea (mullet, Black Sea barbecue); developing the Black Sea as a spawning and feeding area, but wintering or spawning in the Marble and Aegean Seas (pelamida, mackerel, some species of sea karasi) [3].

Of the total number of marine fish inhabiting the Black Sea, 122 are alien from the Mediterranean Sea and 31 are restricted to the Black Sea. Around 20% of them are exploited. The ichthyofauna of the Black Sea, due to contamination of its depths with hydrogen sulfide, is characterized by a greater number of pelagic fish and a limited number of bottom fishes, so that the fisheries are based on pelagic fish. The most important commercial fish are: the Black Sea Sprat and the Black Sea anchovy (Hamsa), which are fish with a short life cycle, feed on zooplankton and have a high reproductive capacity. The number of most Black Sea fish depends not only on the conditions of their existence in the Black Sea, but also on the conditions of spawning, feeding or wintering in the adjacent seas, which determines the complex type of dynamics of the raw material base of the entire sea.

Of the total number of fish, about 20% are exploited. In the 70s and 80s, the USSR caught about 200 thousand tons of fish and seafood in the Black Sea. The basis of the catch was formed by the Black Sea race of anchovy, sprat, whiting, horse mackerel, katran. Catch of other fish—mullet, baraboule, herring, perch (smarids, laurel, luphar, humpback whale) near the coast of the former USSR is very limited due to their low numbers.

Fishery studies have shown that significant interannual fluctuations in the Black Sea fish population are accompanied by changes in the species composition of catches. Thus, from the late 40s to the mid 50s, plankton-evading fish—Hamsa and Black Sea horse mackerel—dominated the Black Sea. Later, until the 1960s, catches were dominated by the Black Sea anchovy (Hamsa). Different researchers estimated the initial stocks and production of fish in the Black Sea at 0.5–5.7 million tones and 0.25–2.9 million tonnes, respectively. Such a large scale is related both to the methodological approach and to large inter-annual fluctuations in the number of commercial fish in the reservoir. In addition, anthropogenic factors are now a significant “regulator” of commercial fish numbers, which affect not only the abiotic but also the biotic part of the Black Sea ecosystem. The results of scientific researches for the last ten years allow us to speak about the initial stock of pelagic fish (anchovy, horse mackerel, sprat) at the level of 2–3 million tons, demersal fish (whiting, catran, stalkan etc.)—0.3 to 0.7 million tons. This estimate does not include data on Mediterranean migrants (Lufar, mackerel, pelamide), as their migration to the area of the former USSR has been practically not observed in the last 20 years.

The category of fishery in the Black Sea includes 15 species of fish, such as flounder (kalkan), mullet, pelengas, anchovy (hamsa), sprat, baraboulka, horse mackerel, etc. Bulls from the family of perches are especially respected. Black Sea sargan also belongs to commercial fish. It is surprising that its bones and blood are green. Katran or Black Sea shark Adult female can reach length up to 2 m and weight up to 15 kg. The male is slightly smaller. There is no direct danger to humans (https://vsya-planeta.ru/flora-i-fauna-chernogo-moria/). Swordfish is a native inhabitant of tropical waters, but periodically appears in the Black Sea and is a large representative of marine fauna: up to 4 m long and about 0.5 tons of weight. Needlefish is a food base for larger marine predators. The appearance is fully consistent with the name. The long narrow body up to 23 cm long is very similar to a needle. It has no gastronomic value. It is a food base for larger marine predators. Locals often use the unusual fish to make souvenirs.

The sea devil Externally very unattractive fish. Has a slightly flattened body on top, covered with outgrowths, and a huge toothpaste. It belongs to the bottom species, lives at a depth of 30–50 m. Spends most of his life "in an ambush" waiting for prey. Very gluttonous and able to swallow fish of the same size as itself (up to 1.5 m long) (https://vsya-planeta.ru/flora-i-fauna-chernogo-moria/).

Scorpena or sea ruff The body of the fish is covered with poisonous spikes, a shot about which is very painful and can lead to intoxication with fever and swelling of the injured area. Fatal for a person meeting with a scorpena is not a threat, but the fish risk their lives. It's being hunted for its delicious meat.

The commercial importance of the Black Sea is determined not only by fish resources, but also by significant reserves of invertebrates (mussels) and algae (phyllophora, cmstuosira, zostera), whose size of populations and associations undergo significant changes under the influence of various economic activities.

Dolphins In addition to fish, invertebrates and algae, the Black Sea is home to mammals. Thus, there are three species of dolphins (whitefish, aphaline and Azovka), which have long been hunted by all the Black Sea countries. The number of dolphins was previously large, and the total production exceeded 10 thousand tons per year, which led to a sharp decline in their stocks. Since 1966, fishing for dolphins is prohibited.

The Afalins are the biggest. The length of the body can reach 3 m and weight—more than 100 kg. To maintain life, they must eat at least 15 kg of fish per day. They are wonderful actors and most often become inhabitants of dolphinariums. Those who doubt whether there are dolphins in the Black Sea, should go for a boat trip closer to sunset and watch the frisky beauties. Afalina belongs to the tooth whales, calves are fed with milk like other species, for which they are classified as mammals. After the pre-mating games 11 months pass until the birth, the babies are not born in eggs, but alive and breathing. The first month they are very restless and practically do not give the mother rest. She looks at it patiently and indulgently and also does not sleep almost a month. The sleep phases of dolphins of all ages are very short. They have to rise to the surface regularly to take a breath of air. So while one hemisphere sleeps, the second works as an alarm clock, time for the next ascent. They live under favorable conditions for about 35 years, the grow up period is long. They communicate with the help of special signals, scientists have counted about 100 combinations, only the most primitive ones, which are accepted by other animals, although the intelligence of these mammals is very high. A total of 4 subspecies of Aphalene are known, including Australian, Indian, Far Eastern and Black Sea. There are about 7 thousand individuals of this species in the Black Sea water area and their number is gradually decreasing. There are several reasons for this: poaching, increasing the number of shipping routes and the intensity of navigation, the remnants of fishing gear in which dolphins are confused and killed, getting into the water harmful substances and oil spillage. In order to preserve the population, the aphalines were listed in the Red Book.

Squirrel. They live in large flocks—the number of individuals can reach 2 thousand. Little White white-fronted Fan of the sea rarely comes to the shore, preferring the depth and freedom. Loves to change his place of residence, lives in the Red, Mediterranean, Caribbean Sea, visits the Gulf of Mexico, seen off the coast of Norway. Regularly appears in the Black and Azov water basins. Body length reaches 2. 4 m, females in the size of a modest, about 1.6 m. Large individuals weigh 100 kg. Predatory body contours, contrasting black and white color and a number of impressive teeth can scare the uninitiated. And completely in vain. The character of the white-white is stricter than that of the aphaline with its known predisposition to humans. But they have no intention of causing harm.

Azovka or porpoise are the smallest and rare enough. Azovka's character is fussy, he prefers to spend the summer in the Azov Sea, where it is warmer, and for the winter goes to the deeper Black. In gastronomic predilections, he also surprises with his legibility. If there is a choice, he will eat only Khamsa. This feature helps fishermen very much, in the direction of the Azovka flock calculating the approximate size and trajectory of the movement of commercial fish. For feeding them enough about 5 kg of fish per day (https://vsya-planeta.ru/flora-i-fauna-chernogo-moria/). Dolphins in the Black Sea are often found. They are protected by law, destruction is prohibited, and the catch is strictly limited. They freely cruise along the straits in the Azov Sea, accompanied by shoals of fish. Dolphins have an amazing ability to echolocation. Species that inhabit the Black Sea, have much in common. They do not fall below 200 m. for food, sleep 10–15 min, while the hemispheres of the brain work alternately.

8.3.2 Features of the Black Sea Fishing

Since coastal waters of the seas are important in the reproduction of the hydrobionts not only of the coastal zone, but also of open waters, the role of particular coastal areas in the reproduction of fishery objects must be defined. If negative impacts of a particular form of fishing on reproduction in important coastal areas are identified, the organization of reproduction sites with the closure of a particular fishing activity at all or for a period of time (fisheries conservation zone) may be recommended. Ultimately, the entire coastal zone can be subdivided into areas that differ in opportunities for industrial fishing, recreational fishing, aquaculture, or other forms of recreation in the water [4].

The general regime for fisheries in the Black Sea is determined by the principles of rational use of fish resources in accordance with the state of stocks of the exploited objects. However, the lack of concerted action in industrial exploitation and biological resources has led to problems in international fisheries management.

In the Black Sea, the shelf area suitable for fish of the coastal complex is about 22% of the total sea area. About 70% of the shelf area falls in the shallow north-western part of the sea, in other areas its length does not exceed 10 km from the shore.

Fishing in the Black Sea has a long history. The city of Kerch was called in ancient times Panticapaeum—the fishing route. Through the Strait of Kerch there is a water exchange between the Black Sea and the Sea of Azov, which has a distributed impact on the Black Sea. In addition to the annual water exchange between the Black Sea and the Sea of Azov through the Strait of Kerch there are active and passive migrations of hydrobionts of both seas. Salt vats and pits in Kerch have been preserved in some places to this day. Hamsa was an important export product in ancient times. Periods of growth and decline experienced fishing in the Black Sea. More than half of the catch in the Black Sea was made up of valuable fish species: pelamide, mackerel, mullette, loufar, large horse mackerel, flounder-kalcanthon until the 1960s. The total catch of the USSR in the Black Sea in 1938–1960 did not exceed 50 thousand tons. In the 70–80s due to the intensification of trawl fishing the catches of boor and sprat increased, amounting to 1988 300 thousand tons. Development of trawl fishing, regulation of river runoff, change of hydrological regime of Bosporus and Kerch Straits and worsening conditions of fish migration through them, eutrophication of the sea and other anthropogenic factors caused radical changes in the state of raw material base. The basis of catches began to be small pelagic species of fish, boor and sprat (up to 80%). Since the late 1980s, the Atlantic *mnemiopsis mnemiopsis leidyi*, a powerful food rival of zooplanktonophages, which had no natural enemies in the Black Sea at the time, has seen a sharp decline in stocks of mass planktophage species. The changes did not affect the stocks of deeper water sprats. In the late 90s, thanks to the introduction of another crest, *beroe ovata,* a consumer of mnemiopsis, the number of pelagic fish species began to grow gradually.

In the second half of the twentieth century, the total catch of fish and other objects of sea fishing of all Black Sea countries reached 600 thousand tons, of which the share of the former USSR was 200–250 thousand tons, including 100150 thousand tons of Ukrainian fishermen.

The peak of mining in the Black Sea occurred in 1980, when the world catch in this reservoir was 850 thousand tons, including Ukrainian fishermen—over 235 thousand tons. Then there was a steady decline in world catches, which by 1996 amounted to 396 thousand tonnes (including 281 thousand tonnes of boorish fish—71% of the total catch). In other words, the decrease in world catches in the Black Sea during this period was more than twofold. The species composition of catches has also changed. So, if up to 50–60 years catches consisted mainly of valuable species: mackerel, pelamide, mullet, horse mackerel, Kambala-Kalkan, herring and sturgeon, then later on up to 90-s and up to now mainly of boor and sprat.

In terms of taxonomic position, the Black Sea Hamsa is one of the subspecies (geographical races) of the European anchovy. In terms of extraction volume it is the most important fishing object in the Black Sea. By its origin Hamsa belongs to the group of Mediterranean universes and, accordingly, is a thermophilic species. The body of the Khamsa is elongated, not forcefully squeezed from its sides. The length of the fish is, on average, about 12 cm.

The Khamsa reproduces practically throughout the Black Sea water area in waters with salt content ranging from 10–12‰ (the Odessa Bay) to 17–18‰ (most of the sea area). Spawning begins in mid-May at 14–15 °C, reaches its maximum intensity in June-July at 20–26 °C and ends by the cone of August. Individual eggs also occur in September. Spawning occurs in the surface horizons of the sea. Individual fecundity of females may exceed 50 thousand eggs. Zooplankton organisms from Copepoda, Cladocera, Cirripedia, Decapoda, Mysidacea larvae, as well as mollusk and worm larvae form the basis of the Khamsa food base. Hamsa juveniles are characterized by a rapid growth rate—by November the average size of the larvae reaches 70–80 mm ([3]; http://elib.rshu.ru/files_books/pdf/21-18.pdf).

Traditional areas of formation of so-called winter clusters of the Black Sea Khamsa are coastal areas of Turkey from Sinop to Rize and the water area adjacent to the Georgian and Abkhaz coasts from Batumi to Sukhumi. It is in these areas of the sea, mainly at a distance of 1–3 miles from the coast, where the Khamsa is actively fishing with purse seine.

The Black Sea Sprat (below the Sprat) is one of the most abundant fish species of the Black Sea, its stock varies in different years in the range of 200–1600 thousand tons. However, since a significant part of the Sprat remains scattered outside the commercial aggregations and there are no appropriate fishing gear for efficient fishing of such Sprat, the stock of this facility is underutilized. The annual withdrawal of sprats usually does not exceed 30% of the commercial sprat stock, with an available withdrawal of up to 44% (40% was withdrawn in 1989 alone). The share of catches by passive fishing gear and the amount seized is currently small. Under favourable conditions, about 1,000 tonnes are produced by commercial seine, which is only a few percent of the annual catch. Most of the latter is in the summer months. Sprat is found throughout the Black Sea. As a cold fish, it prefers to stick to water layers with temperatures of 7–18 °C. The bulk of the Black Sea Sprat spawn from October to March at water temperatures of 6–19 °C [5, 6].

8.3.3 *Black Sea Horse Mackerel*—Trachurus mediterraneus ponticus Aleev

The Black Sea horse mackerel is one of the main commercial fish in the Black Sea. It is distinguished by its two forms—"large" and "small", which are distinguished by a number of features. The most characteristic differences between them are the growth rate and body size. The length of the "shallow" form reaches 20 cm, rarely more, and "large"—up to 55 cm. In the 1940s and 1950s, the number of "coarse" shapes was significant, but later it declined. Nowadays, individuals of the "large" horse mackerel are rare and isolated. There is no unified opinion about the systematic rank of "big" and "small" forms. The largest expert on the Black Sea horse mackerel Yu G. Aleev considered them to be one and the same subspecies. The number of horse mackerel is subject to significant interannual fluctuations. Currently, due to the excessively intensive fishing and the lack of international regulation, the stock of horse mackerel is at a very low level. As a result, there has been little or no domestic specialized fishing since 1987; Turkish catches have also slightly decreased in recent years. Stavrida occurs at water temperatures of 6–25 °C at different salts, but desalinated areas are avoided. As a thermophilic fish, it is active in warm seasons. In summer it holds both near the coast and in the open sea above the layer of temperature jump from the surface to depths of 25–35 m. During this period, it spawn and fatten intensively. From the second half of August it starts to concentrate in the coastal areas of the sea, and in October-December it migrates along the coast to places of wintering. They are located in the coastal waters of Turkey, off the coast of Abkhazia and Georgia and off the southern coast of Crimea. The most dense wintering aggregations, which determines the extraction of the main share of annual horse mackerel catches during this period.

Off the coast of the Crimea, they begin to form in the second half, less often in mid-November, and become mass only at the end of November or even in December at water temperatures around 120 °C (https://delvaneo.ru/; [3]).

Atlantic mackerel (Scomber scombrus) belongs to the family of **Atlantic** perch mackerel. Its maximum body length is 60 cm and its average length is 30 cm. The body is spindle-like. The scales are small. The back is blue-green, with many black, slightly curved stripes. No swim bladder. The mackerel is a pelagic pack heat-loving fish. Predator, finds food on the bottom, covered with vegetation and large rocks. Mackerel lives at a temperature of 8–20 °C, so it is forced to make seasonal migrations along the coasts of America and Europe, as well as between the Marble and Black Seas. These migrations are of a feeding nature (the food of mackerel is small fish and zooplankton) [3].

Considering the peculiarities of biology and ecology of the main commercial fish of the Black Sea, we can conclude that their yield depends not only on the conditions of existence in the Black Sea, but also on the conditions of spawning, feeding or wintering in the adjacent seas—Mediterranean, Marmara and Azov. This determines the particular complexity of assessing and predicting the dynamics of fisheries resources in the Black Sea [7].

The total fish catch in the system of Marmara-Black-Azov Sea, which reached in 1985–1986. 856–906 thousand tons in 1989 fell to 640 thousand tons. Khamsa catch in the northern part of the Black and Azov Seas in 1980–1988 ranged from 240 to 126 thousand tons. Catch of Khamsa in the northern part of the Black and Azov Seas in 1980–1988 fluctuated from 240 to 126 thousand tons, and in 1989 it decreased to 70 thousand tons, catch of horse mackerel fell from 110 to 115 thousand tons to 3 thousand tons, catch of tulle in 1970–1987 was 77–130 thousand tons, and in 1988–1989 it decreased to 36–40 thousand tons. Obviously, the observed catastrophic fall in catches was directly related to the undermining by mnemiopsis of the feeding base and, probably, direct eating of fish larvae during spawning [7].

Analyzing the modern species composition of catches, we can say that up to 90% of the catches in Russia and Ukraine accounts for only three relatively low-value fish species—boor, tulle, sprat.

The modern catch of marine fish in the Black Sea is 17–21 thousand tons in 2009–2011. The total volume of sea fish catch, excluding boorish, whose catch according to the decision of the Russian-Ukrainian Commission on Fishery Issues is carried out at the expense of the total basin volume, for 2012 the forecast was in the amount of 24,669 thousand tons Forecasted catch volumes are undeveloped, mainly due to small pelagic fish species: boorish, sprat, horse mackerel. The main reasons for under-catch are the outdated fleet, absence of purse-sea fishing vessels and fish acceptance and processing bases. Possible increase in production of small pelagic fish species by scientists of FSUE “AzNIIRKh” is estimated at 60 thousand tons.

Nevertheless, fishing is on the rise in some areas. Thus, in 2010 in the zone of East Black Sea State Fishery Protection (Crimea) total catches were 36,117 tons, which is 15% more than in 2009 and 44% more than in 2008. In recent decades, the catches of the Black Sea Sprat in this area ranged from 48,336 tons (2001) to 17,888 tons (2008), the average annual catch was 29,040 tons. The second place is occupied by the catches of the Black Sea boor with annual fluctuations ranging from 1428 tons (2004) to 4987 tons (2010).

Production in 2010 was 4987 tons is the best figure for the last 10 years [8]. The reserves of the Black Sea Khamsa are reviving, but it does not winter near the Crimean coast every year and its winter migration is not stable. At present, users of aquatic living resources have learned to catch it during wintering migration by multideep trawls and the last two years have almost mastered the limit of its extraction. The third place is occupied by the catch of Azov Khamsa with significant inter-annual fluctuations. The 2010 catch was 8709 tons, which is also the best indicator for the last 10 years. On the fourth place for the coast of eastern Crimea is the catch of Black Sea horse mackerel with pronounced interannual fluctuations from 130.5 tons (2001) to 745 tons (2003). The 2010 production was 176.4 tons. The horse mackerel reserves are very unstable, and this is most likely due to its migrations. The stocks of horse mackerel are very unstable and this is most likely due to its migrations. The fish stocks of the bottom complex are much lower than those of pelagic complex and they are mainly caught by coastal fishing.

The main commercial fish species of the coastal fishery are: Black Sea kalka, katran, rays, Azov-Black Sea mullet, pilengas, and from small ones—barabula, as well as a minimum amount of smarid, scorpena, loufar, etc. Results of investigations for the last fifteen years [9, 10] allow to speak about initial stock of pelagic fish (anchovy, horse mackerel, sprat) at the level of 2–3 million tons, demersal fish (whiting, katran, stalkan etc.)—0.3 to 0.7 million tons. This estimate does not include data on Mediterranean migrants (luphal, mackerel, pelamide), as their migration to the zone of the former USSR has been practically not observed in the last 20 years. The high concentration of Hamsa in wintering aggregations provides a good food base for the kalkan, shark crane, beluga, dolphins and seabirds, which are constantly found near the Hamsa jambs.

The contribution of the Black Sea fisheries to the total Russian fish catch is small. The importance of biological resources in the Black Sea is determined, first of all, by its natural and climatic conditions favorable for the organization of year-round rest of the country's population. The high density of population permanently and temporarily living in the region determines the demand for fresh seafood, which is an incentive to develop coastal fishing. Given the limited biological resources of the Black Sea coastal areas and their vulnerability, priorities should be given to their careful and waste-free use, the development of measures aimed at increasing sea productivity, and the organization of fisheries, taking into account physical, geographical, biological and socio-economic factors.

The following should be highlighted as priorities in the Black Sea

- restricting fishing with active fishing gear in coastal waters;
- restoring wallet fishing as a more sustainable way of fishing;
- establishment of coastal enterprises for processing low-value hydrobionts into fishmeal for aquaculture facilities;
- Priority use of passive fishing gear corresponding to the existing raw material base;
- the development of recreational and sport fishing;
- Increase of fishery and fishery resources of the Black Sea basin due to development of the artificial reproduction and commercial sea and fresh water aquaculture taking into account the available world experience, creation of artificial reefs [11].

Researches of "VNIRO" specialists of the coastal water area of the Russian part of the Black Sea with the help of underwater television also showed that starting from the depth of 20–25 m in the areas of trawlers' operation the destruction of the surface layer of the bottom substrate is observed. There are almost no macrobenthos organisms, the substrate is represented by fragments of broken shells of different sizes. Parallel shafts of soil, which are the result of mechanical impact of trawls, are marked, trawl board tracks and lower selection are clearly marked.

As a result of the long-term impact of trawl fishing on bottom biocoenoses at the present time are observed: a decrease in species diversity of ecosystem components, a decrease in water transparency and, consequently, the rise of the lower boundary of the algal belt, the disappearance of many bottom biocoenoses, deterioration of feeding conditions for valuable fish species, a decrease in the level of natural biological self-cleaning of water and, consequently, a deterioration of the sanitary condition of coastal waters [12].

Therefore, despite the significant under-capture of the limits on the catches of boats and sprats, it is necessary to introduce strict restrictions on the work areas for vessels equipped with trawl gear. The entire coastal zone, which is critical to the existence of coastal fish species and largely determines existing biodiversity, should be closed to trawling. Trawl fishing should be shifted to areas of mass concentration of boats and sprats.

However, trawling of these species of fish is economically inefficient, and the boor and sprats from the trawl bag are of poor quality for subsequent processing. The possibility of pelagic trawling due to the higher value of bottom species creates a permanent incentive to disrupt bottom trawling restrictions. It is expedient to restore purse seine fishing of these species with the pouring of catch by fish pumps. In the coastal zone, fishing should be carried out only with passive fishing gear (steel nets, various types of traps, nets), providing minimal impact on the bottom biocoenosis, the possibility of regulating the species and size composition of the objects of fishing by choosing the place and time of the fishing gear installation and through their selective parameters (mesh size, landing coefficient and number of cells). The requirements of ecologically balanced fishing also imply determining the optimal fishing load by the number of passive fishing gear and their stagnation time for the existing fishing areas.

In general, the Black Sea is now characterized by the lowest fish productivity among all Russian commercial seas, which is only 3 g/m^2 per year. Annual production of phytoplankton, zooplankton and zoobenthos is equal to 7620, 711, 660 g/m^2 respectively. The ratio in catches of planktonophages, benthophages and predators is 82:7:11 [8, 10]. In order to improve the current negative situation related both to the fishing crisis and to the general ecological situation in the Black Sea marine areas, it is necessary to consolidate efforts at the level of international nature protection organizations within the framework of the objectives of the Bucharest Convention on Nature Protection of the Black Sea [4].

Black Sea fisheries resources, with their generally significant but highly vulnerable potential, have until recently been of significant importance to the economies of most coastal States. The quantitative and qualitative characteristics of catches are correspond to the oceanological conditions and hydrological regime of individual areas of the basin, the state ownership of specific waters and the degree of development of the fishing industry in individual countries, as well as to the current environmental situation, biological invasions of alien species and the international legal regime of fishing.

In today's environment there is a need to create an effective system of integrated coastal zone management (ICZM), which was reflected in the decision of the UN international conference on environmental protection and sustainable development. To date, about 90 countries are implementing more than 180 programmes on integrated coastal zone management at the international and national levels. The European Commission regards ICZM as a means of preserving coastal zones along with their biodiversity. In large economic projects, social and economic problems are given due place, but environmental protection is a priority. The European States of the North-East Atlantic are focusing management policies on the protection of the marine environment, scientific research on ecosystems, sustainable use of fish stocks, biodiversity conservation, and the development of tourism in coastal areas of countries. Fisheries management should be based on an ecosystem approach, which is "a strategy for the integrated management of land, water and living resources that ensures their conservation and sustainable use…".

It is necessary to emphasize once again the peculiarity of the Black Sea, which has already been mentioned above, that all plants and animals of the Black Sea coast and depths inhabit the upper oxygen layer. Below is a water layer saturated with hydrogen sulphide. There is an opinion that beneath it there may be a world of habitation. And the reason for the spread of such a theory was an underwater river originating in the Bosporus Strait and carrying its waters to the bottom plains. It does not mix with the Black Sea water, so it brings oxygen and food to the depths. The combination of these factors could have triggered the development of deep-sea life forms. But to find out how true this theory is, it is possible only in practical way —by going down to the sea bottom (https://vsya-planeta.ru/flora-i-fauna-chernogo-moria/).

In conclusion, the climatic conditions in the Black Sea basin are extremely favourable for aquaculture development. Aquaculture, with high demand for food products and limited natural resources, is one of the most developing areas of fisheries. Virtually all of the recent growth in world fisheries has been driven by aquaculture. The rapid development of aquaculture began in the 1970s and 1980s. Since then, total annual fish production has increased almost tenfold. Whereas commercial aquaculture accounted for only 3.9% of world catches in 1970, in 2007 it accounted for 43%, or 55.5 million tones (excluding algae), with a total value of US$69 billion. In 2010, the share of farmed fish production exceeded 50% of the global catch. The advantages of this industry are due to the lack of dependence on the variability of the raw material base, lower energy costs than in the case of fishing, proximity of raw material extraction sites to coastal processing complexes, the ability to supply markets with products of stable quality at any time of year. World experience shows that large-scale cultivation of oysters and mussels can be very effective. If natural banks grow mussels to commodity size in 3–4 years, then artificial cultivation with the right selection of the right place will achieve commodity size in 18 months. Output when grown is 2.3 times higher than in natural condition, and the amount of sand in the sash is 1200 times lower. Breeding oysters and mussels requires no forage. The main requirement for their breeding in natural habitats is water purity.

8.4 Conclusions

The peculiarities of distribution of flora and fauna in the Black Sea are considered taking into account its uniqueness—division into oxygen and hydrogen sulphide zones, which no other sea in the world has.

Representatives of the Black Sea fauna, ranging from jellyfishes to mammals, are described in detail.

As representatives of mammals, all species of dolphins living in the Black Sea are described in detail: their physiology, behavior, distribution areas, wintering grounds and reasons for limiting or banning catches.

Special attention is paid to the description of marine fish species of the Black Sea with a detailed description of the features of the Black Sea fishing for fish such as the Black Sea Hamsa, Sprat and horse mackerel of their wintering aggregations, migration. The catch volumes during the periods of rise and fall of the Black Sea fishery are briefly presented.

There are given modern data on the state of the Black Sea fishing in Russia according to the information of the Center of operational forecast of the fishing of "VNIRO" and prospects of the Khamsa fishing in the Black Sea as of 05.12.2018.

The priorities for improving the efficiency of fisheries in the Black Sea, while respecting the general regime of fisheries, taking into account international regulatory objectives, are provided.

The prospects and opportunities for aquaculture development in the Black Sea environment are analysed.

References

1. Vershinin AO (2007) Life of the Black Sea. Kogorta, Krasnodar, 193 p
2. Vulkanov A (1983) Black Sea. Hydrometeoizdat, 408 p
3. Drozdov VV (2002) Many years of Black Sea fisheries resources variability: trends, causes and prospects. Oceanology. Scientists' notes, № 21, pp. 137–154
4. Drozdov VV (2010) Features of the long-term dynamics of the Azov Sea ecosystem under the influence of climatic and anthropogenic factors. Scientific notes 15, ECOLOGY—St. Petersburg: RGMU, pp 155–176
5. Zaitsev YP (2006) Introduction to the ecology of the Black Sea. GEF-UNEP-BSERP, Istanbul-Odessa
6. Novikov NP, Serobaba II (1989) Modern state and prospects of using the Black Sea bioresources in the conditions of anthropogenic influence. In the collection: the South Seas of the USSR: geographical problems of research and development, Geogr. general in the USSR
7. Panov BN, Gubanov EP, Spiridonova EO (2017) Ecology of the sea. Morkniga, 275 p

8. Makoedov AN, Kozhemyako ON (2007) Basics of Russian fishery policy. FSUE "Rybnatsresursy" Publishing House, Moscow, 477 p
9. Lapshin OM, Zhmur NS (1996) Definition of anthropogenic impact on coastal ecosystems and development of a model for balanced management of coastal fisheries. State and prospects of scientific and practical developments in the field of mariculture in Russia: Proceedings of the All-Russian meeting [August 1996, Rostov-on-Don]. AzNIRKh, pp 177–184
10. Luts GI, Dakhno VD, Nadolinsky VP, Rogov SF (2005) Fishing in the Black Sea coastal zone (in Russian). Fishery. № 6, pp 54–56
11. Kumantsov MI, Kuznetsova EN, Lapshin OM (2012) The complex approach to the organization of the Russian fishery on the Black Sea. Modern problems of science and education
12. Boltachev AR (2006) Trawl fishing and its influence on the Black Sea bottom biocoenosis (in Russian). Marine Ecol J 5(3):45–56
13. https://vsya-planeta.ru/flora-i-fauna-chernogo-moria/
14. https://moreprodukt.info/kraby/chernogo-morya
15. https://delvaneo.ru

Chapter 9
Main Natural and Anthropogenic Sources of Pollution of the Black Sea, Its Shelf Zones and Small Water Reservoirs

9.1 Introduction

According to modern estimates, the Black Sea is one of the most polluted areas of the World Ocean. The uniqueness of this semi-isolated body of water from the World Ocean lies in a significant amount of fresh water, 70% of which comes from industrial areas of Europe and contains a large number of different pollutants. Study of migration and time of stay of pollutants in the Black Sea is an actual scientific problem, the decision of which will allow to estimate ability of the sea to self-cleaning [1].

The uniqueness of the Black Sea also lies in the fact that about 80% of the water column and bottom sediments are within the anaerobic zone, they are deprived of other life forms than bacteria, so the sea can be called a bacterial sea. Surrounded by industrially and agriculturally developed countries, the Black Sea serves as a natural lagoon for their runoff. The ratio of catchment land area to its mirror area (422,000 km^2), equal to 4.4, is an order of magnitude higher than the average for the ocean—0.4. The rivers carry into the sea toxic chemicals from fields, pesticides, detergents, limit and aromatic hydrocarbons, heavy metal salts and other toxic substances. Large amounts of fertilizers and organic substances washed off the fields, leading to over-fertilization—eutrophication of the reservoir also goes to the sea, especially in its shallow coastal ecosystems [1].

As a result of significant anthropogenic load on the Black Sea ecosystem, some parts of its water area have lost the ability to self-clean, as evidenced by modern analysis of the environmental situation in the sea. At the same time, the coastal part of the Black Sea is under the greatest pressure, especially in the estuaries of rivers, in the area of ports and the location of large cities.

In addition to river runoff, the ever-increasing stress on coastal areas is caused by stormwater, domestic and industrial runoff from onshore enterprises, communities and recreational complexes. Their local impact is often catastrophic. About 180

K. Pokazeev et al., *Pollution in the Black Sea*, Springer Oceanography,
https://doi.org/10.1007/978-3-030-61895-7_9

thousand vessels with total displacement of 125 million tons pass through the straits of the sea every year, with more than 12 thousand tons of oil entering the sea.

Strong impacts on the biological communities of the sea and its fishing potentials have been caused by irrational, excessive fishing, and in recent years by the shift of new settlers from other areas of the world's oceans, which is destroying the autochthonous communities of the Black Sea (see Chap. 8 Black Sea flora and fauna for more details).

The peculiarities of sea water circulation determine significant differences in the state of the Black Sea coastal and deep water ecosystems (Fig. 9.1). The cyclonic main Black Sea current (MBSc) is traced throughout the sea, creating extensive cyclonic cycles in the central parts of the western and eastern regions (Fig. 9.1) and a number of smaller cyclonic and anticyclonic vortices [1].

In the eastern part of the sea there is a cyclonic cycle, while in the south-eastern part of Batumi there is an intense anticyclonic cycle. In cyclonic cycles, water rises, and in the anticyclonic and outer periphery of the main Black Sea current, it slows down.

In the coastal zone at the edge of the shelf, the anticyclonic swirling of the current occurs as a result of meandering (MBSc) and its lateral friction against the continental slope, where anticyclonic vortexes are generated. In the anticyclonic vortexes, water accumulation and subsequent sinking occur, which coincides with the sinking of water at the periphery of cyclonic cycles. Their addition causes active water immersion near the continental slope and the appearance of a quasi-stationary convergence zone. Convergence at the outer boundary of the MBSc is clearly manifested on the surface by slicks and accumulations of floating debris, especially in calm weather. Off the coast of the Caucasus and Anatolia, the convergence zone is 5–7 miles away from the shore, while in the west and especially in the northwest, along with the edge of the shelf, it goes far away and is expressed weaker [1].

Compensatory upwelling of waters on the mainland slope, which intensifies when anticyclonic vortices pass along the coast, predetermines the existence of a convergence zone beyond the edge of the shelf. The water rising from the cold

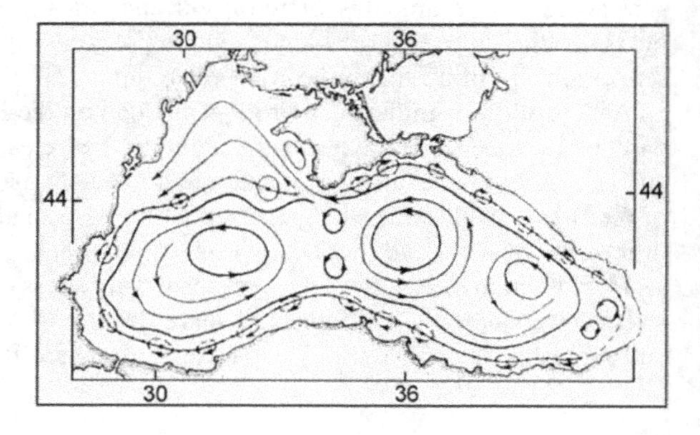

Fig. 9.1 General scheme of the surface circulation of the Black Sea [2]

intermediate layer contributes to a sharp drop in temperature along the coast. A compensatory rise in water may also occur beyond the convergence zone. Thus, a frontal zone similar to the sloping fronts of the open ocean is formed near the Black Sea continental slope. The specifics of water circulation in the Black Sea described above mainly determine the processes of transport, transformation, and disposal of pollutants entering the sea area from various sources.

When heated and distributed coastal waters are mixed with colder and saltier waters of the main Black Sea current, water compaction and sinking occur. Gravitational settling of organic and inorganic suspension as well as turbulent isopic movements result in suspension enrichment of waters along the mainland slope. The coastal waters dropping at convergence carry away from the surface layers to the depth of the pollution falling from the shore and to some extent isolate the open sea waters from the main volume of the coastal anthropogenic runoff [1].

The formation of extremely sharp oxycline in the sea (a sharp drop in oxygen concentration from 5 to 7 ml/l, typical for mixed surface and cold intermediate layers, up to 0.35–0.5 ml/l) is due to the existence of stationary pycnocline that prevents mixing of upper and deep waters. The gradient of oxygen concentration in it can reach 0.2–0.55 ml/l and higher values. Oxycline is formed during oxidation of the dropping dead organic suspension, which is trapped on the pycnocline, and due to rising from the depth of reduced forms of iron, manganese, nitrogen. About 30% is spent on oxidation of organic matter, and about 70% of the total amount of oxygen consumed in the oxycline layer is spent on oxidation of reduced forms of Fe, Mn, N. Deeper oxycline quantity of oxygen continues to decrease and at depth of hydrogen sulfide occurrence (more than 0.005 mg H_2S/l) it still has concentration 0.1–0.3 ml/l [1].

Depending on the interannual differences in runoff intensity, the amount of pollutants entering the Black Sea fluctuates greatly, but remains very high overall. Many countries have taken measures to reduce runoff pollution in recent years, but this has been done poorly and the sea is extremely polluted, especially in its coastal zone.

The structure of this section of the book is such that the analysis of the Black Sea pollution levels in the coastal zone is considered simultaneously with the assessment of its impact on the deep sea zone, since the main sources of anthropogenic pollution are located in the coastal zone of the sea and their impact on the deep sea zone is mainly determined by the specifics of water circulation in the sea. This section analyses current data on the levels of sea pollution by oil products, heavy metals, Municipal solid waste (MSW) and plastics, and radionuclides, as well as estimates of nutrient inputs to the sea as a source of eutrophication of shallow coastal waters and the spread of this phenomenon in the deep sea basin.

This section analyzes current data on methane gas emissions as a natural source of marine pollution in both shallow and deep seas and assesses their impact on marine ecosystems.

9.2 Levels of Oil Contamination in the Coastal and Deep-Water Parts of the Black Sea

Peculiarities of contaminants behavior in the sea area are considered and analyzed taking into account the abovementioned oceanological characteristics of the Black Sea waters.

The most common pollutant in the hydrosphere is oil and petroleum products. Oil is a complex mixture of hydrocarbons and their derivatives; each of these compounds can be considered as a separate toxicant. It contains more than 1000 individual organic substances containing 83–87% carbon, 12–14% hydrogen, 0.5–6% sulphur, 0.02–1.7% nitrogen and 0.005–3.6% oxygen and a small admixture of mineral compounds; oil ash content does not exceed 0.1%. The main characteristics of oil are presented in Table 9.1 [3].

To estimate oil as a polluting substance of the natural environment it is suggested to use the following features: content of light fractions ($t_{boiling}$ < 200 °C); content of paraffins; sulfur content.

The most toxic, in addition to water-soluble phenolic compounds, are volatile aromatic hydrocarbons. Aromatic crude oils (for example, Kuwaiti oils with high sulfur content—up to 2.5%—prevents oxidizing processes) in the initial stage of the spill will be more toxic than highly paraffin Libyan oil with sulfur content of 0.21%. Oxidation can be catalyzed by sunlight and scattered metals such as vanadium, which is present in the oil. Oil slicks reduction and oil emulsification will be facilitated by the formation of water-soluble or surface-active products. The compounds originally represented in crude oil, such as resinous "asphalt" particles

Table 9.1 Characteristics of oil composition and properties

Factions	Mass share in crude oil (%)	Boiling temperature range (°C)	Solubility in distilled water 10^4 (%) (by weight)
Paraffins			
C6–C12	0.1–20	69–230	9.5–0.1
C13–C25	0–10	230–450	0.01–0.004
Cycloparaffins			
C6–C12	5–30	70–230	55–1.0
C13–C23	5–30	230–405	1.0–0
Aromatic hydrocarbons			
Mono- and dicyclic C6–C11	0–5	80–240	1780–0
Polycyclic C12–C18	0–5	240–400	12.5–0
Nafteno aromatic hydrocarbons			
C9–C25	5–30	180–400	1.0–0
Remains	10–70	400	0

and sulphonic acids, may also contribute to the formation and stabilization of emulsions [3].

Lighter fractions have increased toxicity to living organisms, but their high volatility contributes to rapid self-cleaning of the natural environment. Paraffins, on the contrary, do not have a strong toxic effect on soil biota or plankton and benthos of seas and oceans, but due to high curing temperature significantly affect the physical properties of the soil. The degree of danger of hydrogen sulphide pollution of soils and surface waters can be specified by the sulphur content in oil.

When released into the aquatic environment, oil is spilled over the water surface in a thin, often molecular, layer and forms an oil slick that covers tens, hundreds and thousands of square kilometres, depending on the magnitude of the release. Oil hydrocarbons are gradually losing their original individual properties as a result of physical, chemical and biological processes influenced by water and sunlight. Therefore, the introduction of crude oil, its individual components and refinery products into the aquatic environment is generally regarded as a single category of oil pollution. Moving on the surface of the ocean under the influence of wind, currents, tides, oil dissolves, precipitates, undergoes photolysis and biological decomposition. Due to the decomposition and transformation of individual components, the composition of oil is constantly changing. As a result of observations, it has been established that up to 25% of the oil slick disappears within a few days due to evaporation and dissolution of low molecular weight fractions, and aromatic hydrocarbons dissolve faster than paraffins with open chains [3].

Ultraviolet component of solar radiation significantly accelerates the destruction of oil components, but from an environmental point of view this process is dangerous due to the formation of decomposition products, which are usually highly toxic to hydrobionts. After evaporation of the most volatile components the process of oil film destruction slows down as the residues are subject to biological and chemical destruction. Since there is no particular type of microorganism in nature that can destroy all the oil components, the biochemical decomposition of the bulk of the spilled oil is very slow. Bacterial action on the oil components is highly selective, and complete oil decomposition requires the action of numerous bacteria of different species, and the destruction of the resulting intermediate products requires its own microorganisms. Microbiological decomposition of paraffins is the easiest way to proceed. The more persistent cyclo-paraffins and aromatic hydrocarbons last much longer in the ocean environment. The rate of decomposition of hydrocarbons of oil depends on the factors that determine its microbiological activity, i.e. temperature, oxygen access, nutrient regime of the aquatic environment. In water depleted with oxygen the decomposition of oil slows down [3].

The presence of suspended organic particles, bacteria and plankton in water bodies promotes the formation of persistent emulsions as a result of the interaction of heavy oil fractions with sea water. Over time, emulsions coagulate with the formation of resinous clots, which float on the water surface and are released by the tide on land, polluting coasts, beaches and port facilities. The processes of chemical oxidation of oil in the water environment are much slower—their rate is only 10–15% of the rate of biochemical oxidation [3].

Oil was first produced in the Black Sea region on the Kerch Peninsula back in 1864. However, at that time there was no explosive growth of oil production, only small fields emerged, but since then it was known that the Black Sea is rich in "black gold". Full production of oil in the Crimea began only in 1933. Significant resources were invested in the development of oil production, and later gas production, and this has brought results—the production of energy resources today fully covers the needs of the peninsula. Potential revenues from oil and gas production on the peninsula can be very large. According to official sources, Crimea has 47 million tons of oil and 165.3 billion cubic meters of gas reserves. In total, 44 fields may be involved. In the zone of the Kerch Strait, the volume of reserves is projected at over 300 billion cubic meters of gas and about 130 million tons of oil and condensate. The Crimean company chernomorneftegaz (crimea-gaz.ru›chernomorneftegaz/) has been actively involving international partners in its projects for 10 years. The first company to come to the peninsula was Shelton Canada, with which the Crimean monopolist has been working closely since 2006. In recent years, the companies have jointly implemented a project to develop a field located in the western part of the Crimean peninsula on the sea shelf [4].

In most cases, oil pollution at sea is caused by accidents involving ships, bunkering operations at ports and sea trans-shipment points, and accidental industrial emissions. In the Black Sea, oil film pollution is most frequent along the Caucasian coast and near the Crimean peninsula. In open waters, the level of pollution is relatively low, but in coastal waters the maximum permissible pollution rates are often exceeded. According to data [5] in the content of oil hydrocarbons in the coastal waters of the Crimean peninsula in 2016 there are individual cases of exceeding the maximum allowable level of pollution in the surface horizon in autumn. In general, for the Azov-Black Sea water area, the authors note the increased content of oil hydrocarbons in the surface water layer, which is a sign of their preferential entry from land.

Novorossiysk, Tuapse are major oil ports. Almost all oil produced in the Caspian fields comes via pipelines to Novorossiysk, where it is pumped into tankers (Fig. 9.2).

Because of accidents, up to 50 million tons of oil are spilled into the world's oceans every year, and Russia is the second largest oil producer in the world, and has a certain place in the international oil transportation market. Exports of oil products in Russia are mainly carried out through Black Sea ports. About 60 million tons of oil annually leave Novorossiysk by tankers, about 30 million tons—from Tuapse, 3 million tons—from the port Kavkaz (near which the disaster occurred in 2007). In total, more than 138 million tons of oil and oil products pass through the Black Sea ports and arrive at ports not only in Russia but also in Georgia. Following the commissioning (2003) of the Caspian Pipeline Consortium pipeline with an annual capacity of 68 million tons of oil and a terminal in Yuzhnaya Ozereyevka (between Novorossiysk and Anapa), the Black Sea has become the zone of Russia's main oil export [7].

According to the agency Portnews (www.portnews.ru), "23% of all Russian "Black gold", 74% of Kazakh and 65% of Azerbaijani oil exports are exported by the Black Sea. Oil exports through Black Sea ports are very likely to increase in the future".

Fig. 9.2 Oil terminals of the Sheskharis base in Novorossiysk source [6]

Fortunately, there have been no major disasters in the Black Sea, with all the oil spilling out of the wreckage of a giant tanker and covering with film kilometers of sea surface. However, smaller-scale accidents happen. Several ships crashed in the Kerch Strait area during the November 2007 storms, and up to 100 tons of oil products spilled in one place at a time. Smaller accidents taken together add at least as much oil pollution to the Black Sea as the spill in the Kerch Strait in autumn 2007. The wreckage, sinking of any ship, it is always entails a fuel spill, as it is necessary on every ship—for the work of the ship's machine. In a year on the Black Sea passes about 50 thousand ships, each pollutes the sea a little, and in case of accident—strongly, and as a result, in one 2000-th year in the sea was 110 thousand tons of oil products.

With relatively low volumes of accidental spills in the Black Sea, the threat of sea pollution is significantly increased by the construction of new oil storage facilities. The discharge of oil-containing (ballast) water from tankers has a negative impact on the water environment. They provide more than half of the total oil pollution. The reason is the impossibility to fully discharge oil from the tanker. After draining the product, about 0.4% of the transported oil volume remains in the holds. An empty oil tanker loses control due to low draught, so sea water—ballast—is pumped into cargo holds and mixed with oil residues. Before a new loading, the oil-containing water is discharged overboard. Washing of tankers in open waters is also practiced.

For example, in the Tuapse region, oil film sometimes covers up to 90% of the water area. As it follows from (http://литист.рф/экология-новороссийска/черное-море) "oil sources are seagoing ships, which before entering the seaport "cleanse", draining everything superfluous into the sea". Apart from illegal dumping of oil products by sea vessels entering the Novorossiysk seaport pollute the sea, another source of pollution are storm drains, where used oil products enter at night from truck stops that deliver scrap metal for loading to ships in the eastern part of the port, where used oil products are discharged into the storm system. With the first rain, all dirt is washed into the sea and oil stains form on the surface. Another example of oil contamination in the bay is the oil spill through air ducts on tanks during refueling. One of the levers to encourage owners and captains of ships to comply with safety regulations and prevent oil products from entering the water is to conduct inspections on bunkering tankers. According to the Novorossiysk Transport Prosecutor's Office, all fuel spills in the bay are caused by ship crews, which for various reasons allow spills, as a rule, during bunkering and more often at night.

The work [9] presents the results of monitoring the level of oil pollution of the Black Sea coast in the area of Novorossiysk in 2007–2011. The data on the content of hydrocarbons, polycyclic aromatic hydrocarbons and resinous asphaltenes in coastal waters and bottom sediment columns are analyzed. The natural and anthropogenic factors influencing the distribution oil components in the coastal zonehave been revealed. The conclusion about chronic character of oil pollution of the Novorossiysk region coast and its high level is made.

However, in 2011–2013 there was a steady tendency to decrease the level of hydrocarbons content to 0.06 mg/l or less against the background of a certain reduction in the total cargo turnover of commercial ports of the region, including the volumes of oil and oil products shipments due to the crisis phenomena in the world economy. In particular, the total cargo turnover of Novorossiysk port terminals decreased by approximately 8.9% from 2009 to 2013. However, this decline was compensated by 2015 and by the end of 2016 the level of cargo turnover reached 131.4 million tons per year, exceeding the 2009 figure by 6.3%. At the same time, crude oil transshipment in Novorossiysk increased slightly by 0.9% in 2017 compared to the 2016 level [10].

According to data [11] the highest concentrations of oil products during the year are typical for the Black Sea water area in the areas of Novorossiysk and Tuapse (0.015–4.90 mg/l), where there are large oil ports and the most intensive oil transportation by sea. The resort waters of Anapa and Gelendzhik are much cleaner in spring (less than 0.068 mg/l); in summer and autumn the concentration of oil hydrocarbons increases two or more times. Relatively low levels of pollution have been identified in the Republic of Abkhazia, but active shipping, fishing and transboundary transport from Georgia increase pollution levels (to 0.292 mg/l), especially during the spring period.

Oil and its products (gasoline, fuel oil, oils) alsoget into the sea from the shore—with the flow of each river through the soil from the places of ground leaks of oil products; the most significant sources of such pollution are refineries located near the ports. For example, environmental organizations—both public and state—report on the oil lens in the soil created by the refinery complex in Tuapse, which gradually seeps into the sea. Similar reports have been made about the Batumi oil complex and the port in Georgia [12].

Over time, oil products caught in the sea are decomposed by bacteria and life at oil spill sites is restored. After all, oil itself is a natural product, and in the absence of oxygen, the remains of living organisms turn into it. Oil also enters the sea naturally—it oozes out of the oilfields under its bottom, so it is not new to the sea inhabitants. But still, every precaution must be taken to avoid oil accidents; the sea will return to normal after a spill—but it won't be so soon.

As a result of the known oil spill in the Kerch Strait in November 2007. The coastal strip of the Kerch Peninsula facing the Sea of Azov and the adjacent, very shallow water area was mainly affected. However, in summer 2008 the level of MPC of oil was 1.6 more units, and already in 2009 it was below the maximum permissible threshold. That is, it became the same as it was before the tragedy V work [13] analyzes the results of three-year observations of changes in the content and component composition of fuel oil spilled in the Kerch Strait in November 2007 due to the crash of the tanker "Volgoneft-139". The peculiarities of pollutant redistribution between water column, bottom sediments and coast in conditions of small shallow water area are considered. The directions of fuel oil units transportation from the accident area with longshore drift currents are revealed.

For the territory of the Kerch Strait, the diversity of wildlife is low due to over-fertilization of river runoff and low salinity, but the productivity of the marine ecosystem is high. This fact makes it possible to assume that the consequences of the oil spill in 2007 will have a greater impact, firstly, on the lives of local residents and, secondly, on resort revenues. And the local wildlife—not rich, but very beautiful—can recover in 2–3 years, as evidenced by the above information. Powerful flowering of phytoplankton (including toxic species), which captures the entire water area of Azov, and as a result, leading to annual massive fish die-off, have represented and continue to represent a much more serious, environmental problem than the fall oil spill in 2007.

9.3 Contamination of the Black Sea with Ballast Water

Millions of tons of ballast are dumped by sea vessels every year in Black Sea ports. For the port of Novorossiysk this figure is over 32 million tonnes per year. Therefore, the Novorossiysk seaport administration ensures a high level of environmental safety control.

The International Maritime Organization (IMO) is an international intergovernmental organization of the United Nations. IMO activities are aimed at improving the safety of maritime navigation and preventing environmental pollution from ships, primarily marine pollution. In 2004, IMO developed and adopted the text of the Convention on the Control of Water Ballast and Sediments from Ships. The Novorossiysk Seaport Administration, as part of the IMO initiative, has started organizing ballast water control for ships. At the first stage, a temporary methodology for ballast water control on ships was developed and tested, including biological study of species composition of hydrobionts and ecosystem monitoring of biodiversity at ballast discharge sites with assessment of risk of invasive species introduction. The composition, abundance, origin and main distribution vector of invasive species of hydrobionts have been identified, and the risk level for coastal ecosystems in the Azov-Black Sea basin has been assessed. The second stage of the work included a detailed inventory of biological resources and basic biological research in the Novorossiysk Port water area. Sample studies of zooplankton, benthos in control points of the port water area and concentrations of petroleum products were carried out during 2004–2005. Ballast testing continued for density (salinity) indicators. A total of 1308 vessels were inspected. The final phase of the port survey is dedicated to the development of legislation and regulations to prevent the introduction of pathogenic and potentially dangerous biological organisms. Methodological guidelines on ballast water control for ships have been published. According to the decision of the Black Sea states, ballast should be replaced in the Black Sea on the move at depths of more than 200 m. Today, this is the most effective way to protect the Black Sea ecosystem from alien invasive species. Marine biologists call ballast water pollution a “time bomb”.

According to the International Association of Independent Tanker Owners [14] at the beginning of the XXI century the Black Sea was the first in terms of oil products pollution. While in the open part of the Black Sea the level of pollution with oil products is relatively low, in coastal areas, especially near ports, it often exceeds the maximum allowable norms due to planned or emergency discharges from ships as well as from land-based sources.

It should be noted that with relatively small oil spills of about 110 tons per year, the increasing volumes of oil and oil products transportation, construction of new oil terminals increases the threat of major accidents. In case of such an accident, the sensitive ecosystem of the sea may not come back to normal for decades [15].

For the east coast of the Black Sea, the danger is that the main Black Sea current and winds are directed from the oil terminal in Anapa towards the children’s resort, and therefore there is a high probability of accidental oil dumping on Anapa’s beaches. Besides, there are always admissible oil losses during the normal operation of oil ports. According to the experts’ calculations, the Caspian Pipeline Consortium (CPC’s) design materials show such tolerance from 2000 to 4000 thousand tons of oil per year. It should be reminded that one ton of crude oil can cover up to 12 km^2 of water surface with airtight micro-film. The solution of this environmental problem is full and strict implementation of environmental regulations, continuous technological improvement of the equipment of oil pumping and transportation facilities, installation of containment boom barriers on ways of possible oil spill with the purpose of its immediate elimination [15].

High values of oil products content in the surface sea layer are observed in the area of the Southern coast of Crimea (1.33 mg/l) and along the Bulgarian coast (1.25 mg/l). In summer, such areas also include Prykerchenskyi (1.33 mg/l), the south-eastern coast (1.37 mg/l) and the Pribospora region (1.40 mg/l). In the autumn period, the largest deviations continue to be in the Pribospora region (1.14 mg/l), along the eastern and southeastern coast (1.15–2.16 mg/l) and on the routes of sea vessels (1.30–2.16 mg/l) [1].

In the Black Sea, maximum contamination of dissolved and emulsified oil occurs in areas of chronic contamination, where oil products (OP) into bottom sediments of ports and terminals (Table 9.2).

Table 9.2 OP concentration in sea bottom sediments of main ports of Ukraine and Sevastopol [16]

Concentration of oil products (g/kg)	Ports				
	Odessa	Ust'-Dunai	Ilyichevsk	South	Sevastopol
	3.8	4.5	2.7	1.9	12.8

According to data [16] the highest level of pollution is registered in bottom sediments of ports, especially Sevastopol (Table 9.2). The authors believe: «Given the political and economic situation in the Crimea since March 2014, the attempt to blockade the Crimea by Ukraine, which led to a sharp increase in the flow of passenger and cargo transportation by sea, increasing the intensity of shipping and port use, the situation with oil pollution of the Black Sea off the coast of Crimea will only get worse».

Further analysis of the contribution of the Black Sea states to the Black Sea ecosystem pollution has shown that the most significant pollution of the sea area belongs to Ukraine (65.8%), followed by Romania (12.7%), Bulgaria (11.8%), Russia (7.8%), Turkey (1.6%) and Georgia (0.3%) [17].

The main bulk of observations of oil hydrocarbons (OH) in the Black Sea, in its coastal waters in the last century was made by marine units of the former USSR State Hydrometeorological Committee, Danube GMO, MeteoHydroStation (MHS) "Odessa", Nikolaev GMO, GMB "Ilyichevsk", MHS "Yalta", MHS "Tuapse" and Adzhar GMO. An array of observations of coastal waters of the Sevastopol region and the open water area of the sea was obtained Sevastopol Branch of the State Oceanographic Institute (SBSOI) and then Marine Department of the Ukrainian Scientific Institute of Hydrometeorological Research (MD Ukr SIHMR) and is fully described in the work [18]. Yearbooks on sea water quality on hydrochemical indicators are being produced. The work is carried out in the Laboratory of Marine Pollution Monitoring of the State Oceanographic Institute of Roshydromet (LMPM SOIn), Moscow, www.oceanography.ru, section "Pollution of the seas"). The Yearbook contains average and maximum for a year or a season values of individual hydrologic and hydrochemical indicators of sea waters of controlled coastal areas, as well as the level of pollution of water and bottom sediments by heavy metals, oil products and a wide range of organic substances of natural and anthropogenic origin [19].

According to the data [19] in the Sevastopol Bay and on the seashore of 117 samples taken for analysis of oil hydrocarbons, in sixteen of them the concentration was below the detection limit (Detection Limit DL = 0.01 mg/dm^3). In June, the water pollution in the bay varied within 0.01–0.07 mg/dm^3, with an average of 0.039 mg/dm^3. In four of the nine samples taken, the oil hydrocarbons (OH) concentration was equal to or exceeded the maximum allowable concentration (MAC). The maximum concentration was observed in the surface layer at the bay outlet. At the seaside of the city in Sevastopol, maximum concentrations of OH reached 0.15 and 0.39 mg/dm^3 (3 and 8 MACs) in bottom waters in May and 0.20 mg/dm^3 (4 MACs) in surface waters in June.

In the waters of the port of Yalta oil hydrocarbons were found throughout 2017 in all 72 samples, but their content remained very low. The maximum concentration (0.03 mg/dm^3, 0.6 MAC) was noted at the end of February in both water layers (surface and bottom). In 52 samples, the NO concentration was equal to the detection limit (DL = 0.013 mg/dm^3). The average concentration was 0.013 mg/dm^3 [19].

In the waters of the eastern Black Sea coast near the city of Anapa oil hydrocarbons were found in 24 samples of 27 (DL = 0.02 mg/dm^3). The concentration of DL exceeded MAC in only one sample taken on January 25 (0.054 mg/dm^3). The average annual value changed insignificantly compared to the previous year and was 0.014 mg/dm^3. Oil hydrocarbon pollution in the Tsemesskaya Bay was low: the concentration of MAC in two of the 22 samples treated was below the detection limit (DL = 0.001 mg/dm^3), while in the others it was not higher than 0.028 mg/dm^3, i.e. did not rise above the MAC level (MAC = 0.05 mg/dm^3). The water area of the Novorossiysk port has been showing a tendency to decrease in OH content for three years. In 2017, the average annual concentration decreased by 1.4 times compared to 2016 and by 2.6 times compared to 2015, amounting to 0.014 mg/dm^3. Maximum concentration in 2017 was 2.3 times less than in 2016 and 3.1 times less than in 2015 [19]. Since 2014, it is possible to note a decrease in the average and maximum values of OH in the waters of the Caucasian coast to a greater extent due to the contribution of coastal waters of Tupase. During this period, the average annual concentration in the Tuapse water area has decreased by a factor of three and the maximum by a factor of 14.6.

In 2017, in the coastal waters between Adler and Sochi, the level of **petroleum** hydrocarbons varied from analytical zero (69%—44 samples out of 69) to 57 mkg/dm^3 (1.14 MAC). Last year only three samples from 64 had the content of oil hydrocarbons equal to analytical zero (4.7%), in contrast to 72% the year before last. The average annual value was 10 mkg/dm^3, which is 1.7 times less than in 2016. The maximum value was almost two times less than last year's and was recorded on March 22 in the bottom layer at a depth of 6 m at the mouth of the Khosta River. The average concentration of OH in surface and deep water varied slightly (9 and 10 μg/dm^3 respectively). The waters in the Sochi port area were the most contaminated with OH during the observation periods (18 μg/dm^3 on average), slightly less in estuarine areas (11) and the lowest concentration at 2 nautical miles from the shore (6). In general, over the whole water area of the coastal area of the Greater Sochi, the content of oil hydrocarbons has been gradually declining over the last decade and a half to the level preceding 2003.

However, during May 2020, residents of Anapa, Sochi and Novorossiysk signaled that dolphins were dying due to water and beach pollution with oil products. "We constantly find dead dolphins on the shore, some corpses are completely black from oil film. Over the past month, volunteers of our Center found about a hundred bodies on the beaches of Anapa, Sochi and Novorossiysk said a volunteer of the Center "Dolphin" Tatiana Semenova.

Employees of the Environmental Watch for the North Caucasus appealed to the administration of the Black Sea seaports to clarify the situation. Representatives of the port administration responded that "during the mentioned period there were no cases of oil spills in the water areas of the seaports of Taman, Anapa and Gelendzhik, no cases of discharges from vessels of oil-containing waters on the approach routes to them were registered". As an explanation for the discovery of oil products on the sea coast in several settlements, the port administration suggested "that oil products may have been released as part of storm water from the coastal outlets. Another possible source is the discharge) into the coastal zone of oily pellets from sources of natural origin due to the activity of underwater mud volcanoes along the Tuapse geological deflection and the Shatskiy Valley".

Currently, many mathematical models have been developed to predict the spread of the oil slick after a spill. MGI RAS has also developed an operational system for forecasting the spread of oil spills in the Black Sea [Black Sea Track Web (BSTW)] based on the synthesis of modules of the Baltic Operational System for Oil Spill Forecasting adapted to the physical and geographical conditions of the Black Sea and the operational model of the Black Sea circulation of MGI [20].

The system (BSTW) builds on the latest achievements of operational oceanography and aims to improve the environmental safety of the Black Sea. Examples of the use of the system (BSTW) in real conditions are known and good results have been obtained predicting the spread of real oil spills in the Black Sea [21]. There is data on the use of this system as a scientific substantiation of claims to economic entities that caused some or other technological violations with negative environmental consequences [22].

9.4 Black Sea Pollution by Sewage (Level of Eutrophication of Coastal Shallow Water Ecosystems in the Black Sea)

As a closed basin, the Black Sea collects all pollution from the runoff of the Danube, Dnieper, Southern Bug, Dniester and other rivers. It is with the river runoff that more than 60% of the waste from the territory of 20 countries of industrial Europe comes. Together with fresh water, waste from sewage of coastal cities and the results of the operation of industrial enterprises enter the water area. For example, in areas with developed viticulture an increased content of copper compounds in coastal waters is sometimes recorded.

Biogenic substances are mainly transported by the Danube River, but also by other rivers whose waters flow to the north-western shelf. The oversaturation of water with nutrients and phosphorus causes a process called eutrophication (literally "overfeeding"). Algae, both unicellular and multicellular, begin to reproduce intensively in such water. This leads to the fact that plants use almost all the water-soluble oxygen for their own breathing and do not have time to cover the deficit due to photosynthesis. Animals that are left without oxygen just suffocate.

Thus, eutrophication of water bodies leads not only to deterioration of recreational qualities of the water area, but also to degradation of natural communities [15].

As a result of excess nutrients (phosphate and nitrates) entering the sea, the water blooms. In areas of deep effluent, blooms may also occur when the nutrient concentration in the water increases. As a result, algae multiply rapidly to form a dense layer. The sun's rays cannot penetrate the algae that are dying out. The lack of oxygen at the bottom results in the death of deep dwellers. Bacteria are growing in the water and consuming oxygen. Lack of oxygen in the latter leads to mass death of fish, squid, mussels and oysters.

Not having time to consume and mineralize in the communities of surface layers, the organic suspension coming from the river runoff, it dies out and sinks to the bottom. In the shallow north-western part of the sea, a sharp seasonal thermocline with temperature gradients of up to 15 °C is formed in summer in high-water and low-windy years (this is the regime that has characterized the last decades). Such thermocline formed at depths of 5–15 m blocks the exchange between surface and deeper waters and is not destroyed even in storm winds of 15–17 m/s, blowing for 2–3 days. Due to this, in the bottom layers (up to the depth of 20–40 m), enriched with suspended organic matter, available oxygen is quickly consumed, in vast areas of shallow water deep hypoxia occurs and hydrogen sulfide appears. Usually hypoxia begins in May and continues until September, capturing deeper and deeper parts of the northwestern shelf [1].

The waters of the Danube River, captured by the main Black Sea current, move along the western coast of the sea to the south with the Rumelian Current. Their influence can be traced up to 43° and even 41 °N, and then, with an extensive anticyclonic vortex, they again partly reach the north almost to the Bay of Varna. These waters carry a mass of toxicants, dissolved and suspended organic matter and biogenic elements that have not yet settled on the hydrofronts and have not been used by the communities of the north-western shelf. They cause contamination and eutrophication of the western branch of the main Black Sea current. They are associated with summer blooms of peridinium algae and other negative phenomena that degrade coastal water quality, causing enormous damage to the recreational complexes of Romania and Bulgaria. The influence of desalinated Danube waters can be traced at quite a considerable distance from the shore—40–50 miles, while coastal waters are much more affected by local runoff [1].

In coastal waters, the constant year-round release of high concentrations of organic matter leads to the formation of two stable zones of "increased concentration of life". The first zone extends from the water cut to depths of 5–10 m, the second zone extends to depths of 80–100 m directly above the depth dump. In these zones high rate of production and destruction processes is provided by high activity of enzymes (hydrolases, alkaline phosphotase, nucleoprogeases and redox enzymes), which are responsible for different types of organic matter conversion. Acceleration of organic matter and nutrients turnover in the production and destruction cycle of coastal waters leads to the development of powerful microalgae blooms. These blooms are most pronounced not when there are huge amounts of phosphorus, nitrates, and silicon in the water, as was noted in the Varna Gulf (phosphates—23.5; nitrates—273; silicon—62.3 μg/l), and not when the water contains large amounts of organic matter (Dnieper estuaries), but only when microheterotrophs are well developed and the rate of production and destruction processes is very high [1].

The situation related to the disturbance of coastal water ecosystems in the areas of Black Sea cities has become more acute. It is primarily related to local discharges of domestic untreated or poorly treated wastewater. The most heavily polluted bays and bays of large cities are Sevastopol, Yalta, Novorossiysk, Gelendzhik, Sukhumi, Poti, Batumi and Pitsunda districts [23, 24].

The study of hypoxia arising in coastal ecosystems at the Marine Hydrophysical Institute was carried out using methods of mathematical modeling [25, 26]. It should be noted that the study of hypoxia on the northwestern shelf, in addition to nature conservation, is also of scientific fundamental importance, because in a very short period (about three months) there is a change of oxidation conditions to reducing ones and vice versa, demonstrating a time-limited model of the geological past of the Earth, when the "sulphate" and "nitrate" breath of microorganisms was replaced by "oxygen" and back. Since the processes accompanying hypoxia are similar to those occurring in the contact zone of aerobic and anaerobic waters in the Black Sea, the experience of mathematical modeling of the hydrogen sulfide ecosystem [25, 26] was used in modeling hypoxia on the northwestern shelf.

To model hypoxia, the measured seasonal course of primary production for the shelf zone was used, which is characterized by the presence of a single summer-autumn maximum, in which the value of primary production reaches 0.6 gC/(m^2-days). The flow of organic substance of allochthon origin, which was supplied from flooded river waters at its maximum content (0.1 gC/(m^2-days), was also taken into account.

It should be noted that according to the results of calculations hypoxia occurs at depths of 30–40 m (at a distance of 30–35 km from the shore), then spreads to the shore up to the shoreline of the water cut. According to the calculation results the maximum development of hypoxia is observed on 8–10 days. The oxygen deficit covers a 10-m bottom layer. Sulfate production in the water column is increasing, hydrogen sulfide has no time to oxidize and accumulates at the bottom. The dead bottom fauna and flora itself becomes a hearth of sulfate reduction, closing the positive feedback loop. The following maximum concentrations are typical for this stage of the process: thiosulfates 4 mg/l; molecular sulfur 2 mg/l; thionic bacteria 0.6 mg/l. Hydrogen sulfide production at the bottom and in the water column reaches 4 g/m^2 per day. By the end of suffocation the size of sulfate reduction at the bottom is reduced, the aerobic mode of the water column is restored. It should be stressed that in the calculations the dynamic parameters did not change: vertical exchange was not increased, which in real conditions increases in the autumn period because of the surface layer dewatering and wind wave. Under the model conditions, the oxygen regime was restored after 90 days, the concentration of thiosulphates was reduced to background values after 25 days, molecular sulfur and thionic bacteria—after 30 days. The biomass of benthic organisms was restored more slowly. In general, throughout the entire area of the shelf they were restored only after 450 days [25, 26].

The cause-and-effect pattern of hypoxia and the appearance of hydrogen sulfide lenses on the northwestern shelf of the sea, laid down in the model, is shown in Fig. 9.3.

Thus, results of the executed calculations allow to assume, that the main reason causing hypoxia is coincidence of time of existence of the blocking layer interfering vertical water exchange, with a seasonal maximum of a stream of organic matter on a bottom. Autumn convective stirring certainly accelerates the process of disappearance of the hypoxia focus, but does not change the direction of the process. Factors such as an increase in the rate of all biochemical reactions with increasing temperature, which leads to an effective increase in the effect of the barrier layer, as well as the decrease in oxygen solubility associated with increasing temperature, seem to be of secondary importance. In addition, the results of the performed calculation, namely, a very long time of biomass recovery in bottom organisms, may serve as a sign of the observed in recent years impact of hypoxia on biodiversity reduction in the northwestern shelf [1].

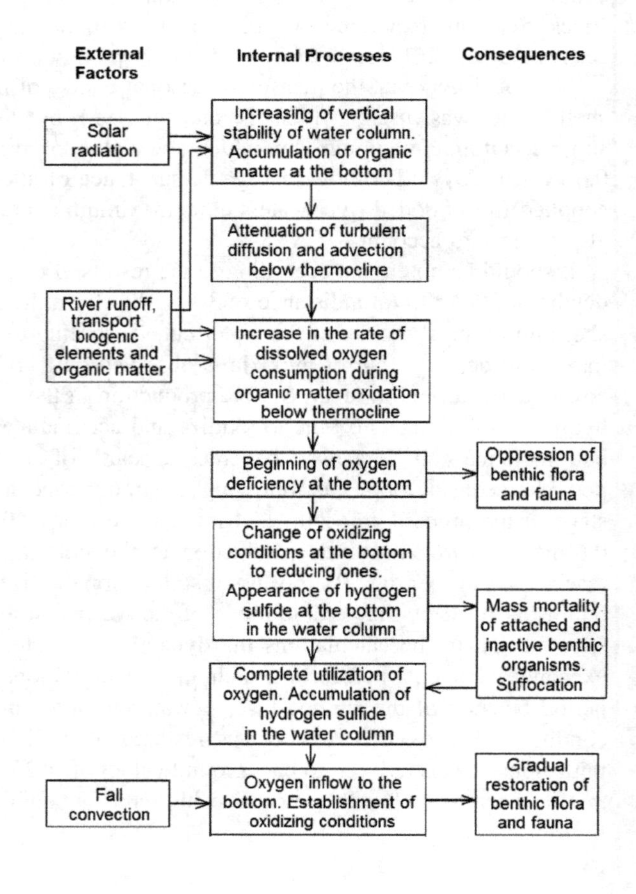

Fig. 9.3 Causal pattern of bottom hypoxia

Another factor in the negative impact of hypoxia on the eutrophication of the north-western shelf is its amplification through the intake of biogenic nitrogen and phosphorus from bottom sediments, the so-called "secondary" eutrophication. According to the work [27], eutrophication on the Black Sea shelf in the period 1996–2000 developed mainly due to a significant increase in the inflow of regenerated nutrients from bottom sediments into the water column after regular periods of hypoxia and suffocation. This phenomenon, driven by both anthropogenic and natural factors, has led to a further increase in primary production on the shelf, with the flow of nitrates and phosphates into the sea from river discharge stabilising in recent years.

Coastal waters of the Southern Coast of Crimea (SCC), including waters of the Yalta Bay, are recreational and resort areas experiencing significant anthropogenic stress with a pronounced seasonal maximum. In addition to anthropogenic load, surface waters of the study area are also affected by a number of natural and climatic factors, such as coastal upwelling, intensive coastal currents, storm runoff and precipitation, which to some extent affect the distribution of a number of hydrological and hydrochemical parameters.

The work [28] assessed the dynamics of biogenic elements in the surface and bottom layers of the waters of the Yalta Bay in all seasons of the year for the ten-year period 2003–2013 against the background of changes in temperature, salinity and dissolved oxygen to determine the impact of resort and recreational activities on the hydrochemical regime of the studied water area. It has been shown that the season of maximum recreational load on the Yalta Bay: June, July, August according to the analysis of a number of hydrochemical parameters indicates certain changes unfavorable for the ecosystem and requires more frequent observations in the summer period with the involvement of hydrometeorological observations and data on the sanitary and epidemiological situation along the coast with the assessment of discharges.

As a continuation of the research in the work [29] the distribution of biogenic elements (total nitrogen, nitrate, nitrite, ammonium, total and phosphate phosphorus, silicon) and dissolved oxygen in the surface and bottom waters of the water area of Yalta seaport for the period of 2013–2017 is studied and the current ecological state of the water area is assessed when compared with the data of 2003–2013. Both in terms of average and maximum values for the period 2013–2017 there is an increase in the total nitrogen content in the water area of Yalta port, which occurs against the background of a decrease in the content of inorganic forms (ammonium and nitrate nitrogen) in contrast to the previous period.

The relative content of dissolved **oxygen** was 94–96% of saturation on the average for a year, changing on the surface horizon within 74–127% of saturation, on the bottom horizon within 76–128%. According to the monthly average values, saturation of water with oxygen only in May and June exceeded 100% of the norm, in autumn-winter period the deficit of dissolved oxygen in the surface-bottom layer reached 7–9% of saturation (Fig. 9.4). It should be noted that in the warm period (April–October), surface water aeration is 1–4% worse than bottom water, which is apparently due to the influence of bottom currents. In terms of absolute values, the

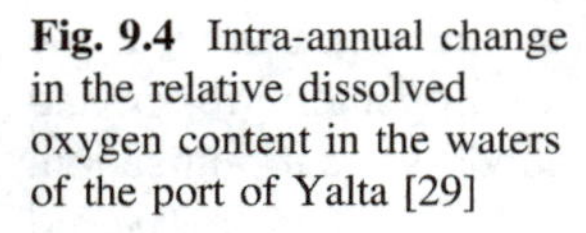

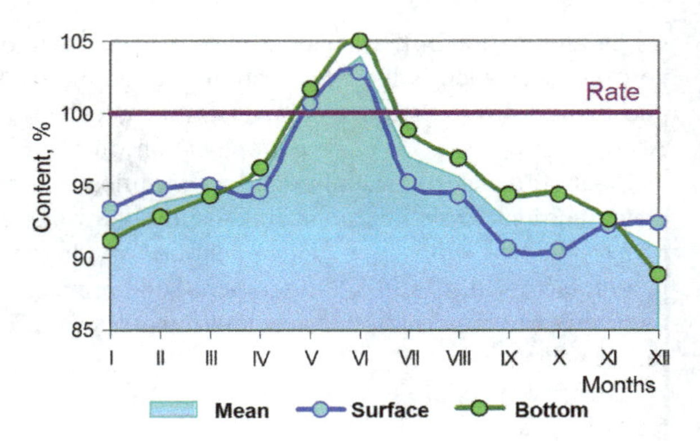

Fig. 9.4 Intra-annual change in the relative dissolved oxygen content in the waters of the port of Yalta [29]

dissolved oxygen content varied between 5.74–12.25 mg/l at the surface and 6.44–11.40 mg/l at the bottom [29] (Fig. 9.4).

Increased concentrations of nutrients are also observed on the Eastern Black Sea coast, especially in port waters, at stations located in close proximity to storm sewer outlets.

The water area of the Novorossiysk port is a breakwater enclosed part of the Tsemesskaya Bay with a complicated coastline with deep pockets. The port's water area is under maximum anthropogenic load under conditions of difficult water exchange with the open sea. As a result, port waters throughout the year are characterized by low transparency due to both general pollution and intensive soil disturbance from shipping and prevailing north and northeast winds. Bottom sediments of the port are of liquid consistency and consist of black silts with a hydrogen sulphide smell. It has been revealed [30] that accumulation of lability sulfides toxic for fauna and recovery sediments with the content of more than 500 mg/dm^3 of raw sludge in Novorossiysk and Tuapse ports are dangerous ecological consequences of anthropogenic pollution causing degradation of bottom biocoenosis. There is also a lack of typical overgrown fauna in the berth area. According to scientists, all of this is not only a consequence of the Novorossiysk Commercial Seaport (NCSP), which is one of the ten largest European ports in terms of cargo turnover, but also an impact of the city, as the port waters take up 65% of the total amount of Novorossiysk's coastal runoff.

"The main source of water pollution in the Novorossiysk port is the coastal runoff: small rivers, storm and sewage runoff," says Valery Chasovnikov, Ph.D. in Geography, acting head of the Laboratory of Sea Chemistry of the Southern Branch of the Institute of Oceanology.—"According to the results of our research, in the areas of loading mineral fertilizers in the port there is no increase in the background of nutrient elements, which would certainly affect the results in the case of fertilizers in sea water. Industrial and citywide releases of water flow into the bay, in some cases without preliminary purification, which, in total, gives the final flow of nutrients and contaminants into sea water".

The significant impact of coastal runoff is evidenced by lower salinity values and higher alkalinity and silicon values in the port water area. The most typical impact of small rivers and stormwater runoff is an increase in nutrient concentrations (gross phosphorus and nitrogen, ammonium, nitrates, nitrites, urea) directly at coastal stations at the water's edge. The intensive inflow of mineral and organic forms of nitrogen and phosphorus is accompanied by increased photosynthesis processes and the creation of a large number of primary products, which may eventually lead to the oversaturation of water with biogenic elements and cause disturbances in the natural balance of the ecosystem.

The first alarming symptoms of this process are high values of biochemical oxygen consumption. Another indicator of active redox processes is nitrite nitrogen, relatively high concentrations of which are also regularly observed in samples. At the same time, it is noted that the hydrochemical regime of the port water area currently meets the established norms. However, the danger lies in the fact that after exceeding the limits of natural self-cleaning, these processes may develop avalanche-like and eventually lead to suffocation [31].

9.5 Contamination of the Black Sea with Municipal Solid Waste and Microplastics

At present, diving research shows that the bottom of the Black Sea coastal waters is literally swamped with household garbage. These are bottles, cans, plastic waste, scraps of fishing nets and much more. A similar situation can be observed everywhere and on the shore. Municipal solid waste (MSW) and plastic has been a real coastal scourge in recent years (Fig. 9.5).

Recently, the problem of marine littering with waste has also become increasingly acute. We see underwater dumps all around; a lot of garbage is brought by the spring floods and rivers. Often landfills are made along river banks; during floods the waste is washed away and enters the sea. During the holiday season, a large amount of MSW is accumulated on the shores of the sea, as public services are not able to cope with the cleaning and, in addition, the system of MSW processing is not properly developed [8].

In seawater, household waste can decompose for years, decades, and plastic waste for centuries. Toxic decomposition products enter the water. Floating plastic debris can be mistaken for food and swallowed by marine animals), often resulting in animal death [32].

In Gelendzhik, scientists take samples of bottom sediments with difficulty, as the "garbage belt" extends 300 m from the shore, and we have to literally punch through almost half a meter of compressed junk. Unauthorized emissions from sewage collectors still occur periodically in the Novorossiysk water area. The situation has slightly improved in the Adler region, where the sewage treatment plant has recently been renovated. Local improvements in sea water have been brought about by penalties imposed on a number of enterprises. A positive example of many years of efforts by environmentalists is the cessation of open coal overloading at the

Fig. 9.5 Pollution of beaches with plastic [15]

Tuapse cargo port. However, in general, a significant part of the Russian Black Sea and Azov Sea coasts today have an 'adverse water' index, and this problem needs to be addressed on a national scale. Even now, many Black Sea beaches do not meet international sanitary standards—and not only because of the pollution of the coasts, but also because of the increased content of pathogenic germs and bacteria in the water. The main contaminant here is domestic sewage, i.e., simply speaking, sewage drains are only two conditions: insufficiently treated and untreated. Further uncoordinated human behavior and thoughtless build-up of the anthropogenic pressure may lead to complete disruption of the ecosystem function of the shelf. Of course, the future of the latter depends not only on the six Black Sea countries, but also on what environmental policy will be implemented by their immediate neighbours. Employees of the Marine Biology Laboratory of the Southern Scientific Center of the Russian Academy of Sciences and the Murmansk Marine Biological Institute together with the Laboratory of Marine Chemistry of the Southern Branch of the Institute of Oceanology of the Russian Academy of Sciences and ecologists of the Federal State Institution "Novorossiysk Seaport Administration" conducted environmental monitoring of marine biological resources in the water use zone of Novorossiysk Commercial Seaport.

The main conclusion made by scientists on the basis of the research was that "the water area is most polluted by urban sewage" (http://литист.рф/экология-новороссийска/черное-море/).

Within the framework of monthly monitoring of qualitative-quantitative distribution of microplastics in soils of beaches of Sevastopol, in work [33] results of quantitative indicators of the content of microplastics in soils of two popular beaches Omega and Uchkuevka in the summer period (May–September) 2016–2017 are presented.

The highest values on both beaches were registered in May 2017 (6.9 ± 0.26 units-m^{-2}) on Omega beach and in August 2016 (3.5 ± 0.088 units-m^{-2}) on Uchkuevka beach. Staying in the sand or pebbles on the sea shore, plastic particles are shredded to even smaller sizes over time. Aging of plastic materials occurs under the influence of heat, ultraviolet rays, air oxygen, water, mechanical influence.

All of the above conditions for the degradation of plastic materials on the beaches are in abundance, so mesoplastics (plastic particles smaller than 5–25 mm [34] and "coarse microplastics" (1–5 mm) are sooner or later crushed to "fine microplastics" (20 μm–1 mm) [35]. Compared with macroplastics (20–100 mm and more), which are regularly removed from beaches during cleaning, microplastics are not visible in the sand or pebbles, and the threat is the probability of its long-term accumulation in the soil beaches. The author [33] connects higher rates of microplastics content on the Omega beach with the strong closure of the Omega Bay, difficult water exchange and gradual from year to year accumulation of microplastics (https://elibrary.ru/item.asp?id=35128533).

The level of microplastics content in the Sevastopol bay and adjacent water area reaches 750 μg per cubic meter. These numbers were presented by scientists of the Federal Research Center "Institute of Biology of Southern Seas named after A.O. Kovalevsky RAS". If we compare the content of microplastics in coastal waters of the Black Sea with other seas and water areas of the World Ocean, we can talk about moderate pollution. Scientists suggest that the situation in the deep Black Sea should be more favorable (Fig. 9.6).

According to Russian environmentalists, the main problems with the use of plastic and the scale of degradation of plastic waste entering the seas are as follows:

- By 2050, plastic production will require 20% of the global oil consumption.
- Every minute a truckload of plastic trash gets into the open sea.
- Each year, disposable plastic causes the death of 100 thousand marine mammals and 1 million seabirds.
- Around 40% of all plastic produced in Europe is packaging.
- Each Russian uses approximately 181 plastic bags per year.
- Already 127 countries regulate the turnover of disposable plastic products.
- More than 4000 chemicals are used in plastic packaging, 148 of which have already been recognized as dangerous to humans and nature.
- More than 90% of all plastic produced has never been recycled.

"Zero Waste" project of Greenpeace in Russia //greenpeace.ru/projects/zero-waste/in November 2018 conducted an expedition on the shores of the Black and Azov Seas, which audited the plastic waste, determined its morphology, that is, the composition of the plastic waste, which is the main pollutant of the water area of the seas and coastal areas. It has been shown that the share of disposable

Fig. 9.6 Plastic trash in the Black Sea [36]

plastic goods is very high: for the Black Sea it is about 70 percent of all plastic goods, for the Azov Sea it is 90%. These are plastic bottles, bags, food packaging, all kinds of disposable tableware, wet wipes, drinking straws. In fact, this is a list of items that are currently being withdrawn from circulation in progressive countries. There are wastes that get to Russian shores from foreign, neighboring countries through transboundary movement. A lot of wastes of Turkish origin, something from Georgia, from Ukraine have been found. It can be assumed that the situation on the shores of those countries is the same, some garbage from Russia can be found there. In fact, this problem has no borders. Of course, ships also leave a certain trace. There is a certain share of marine debris from ships—all kinds of nets, ropes, fishing waste.

The danger of this waste is that pieces of plastic ranging in size from 2.5 cm to a few microns migrate very easily along food chains and have negative physiological effects on animals. There are a huge number of examples of dolphins and whales around the world dying with plastic in their stomachs. Moreover, plastic contains so-called plasticizers, additives, dyes, and these can already be toxic, carcinogenic substances that also migrate in the environment and damage biodiversity. Decomposing over hundreds of years, disposable plastic contaminates the environment and harms sea animals and birds, which often become victims by mistaking plastic fragments for food or getting tangled up in it.

The most significant impact of human activities on the state of the seawater is observed on the coastal area from Anapa to Tuapse. To the south, from Sochi to Sukhumi, the sea is considered “conditionally clean”. However, this "conditionality" also balances on the dangerous edge. As the raids with the use of special helicopters have shown, not only along the portcities but also along all resort towns without exception there are underwater "ridges" consisting of flooded debris, mainly plastic and rubber products. In the Sochi area, for example, this loop stretches annually for 20–25 km with an average width of 800 m. After winter storms, all this mass drifts to the beach water areas. There is a rapid increase in marine pollution by ships, among which in recent years many private ones have appeared. The situation with plastic garbage in Sochi is only getting worse as the number of consumers is increasing every year. Not only the sea is getting polluted, but also river floodplains. A large number of rivers flow into the Black Sea. There are large construction sites along them, a lot of catering establishments and picnic areas. When there is heavy rainfall, it all washes away into the sea. Then storms bring it all ashore or it stays in the sea. Tuapse region is a resort place, and the theme of plastic waste and garbage is very relevant [8].

According to the results of the "Zero Waste" Greenpeace expedition in 2018, plastic pollution is a real problem for the Black and Azov Seas.

Greenpeace experts on the expedition used a technique for monitoring sea debris on the beaches, which was developed by the DeFishGear project. The data collected with its help formed the basis for the European Commission decision to ban certain types of plastic products. Between 435 and 3501 plastic fragments—disposable goods, containers and packaging—were found on the 100-m stretch of coast: items that should not have been there. In Russia today, unfortunately, there is no control over plastic pollution, as well as no systematic measures to prevent it. Greenpeace calls on the Russian government to develop a federal system for monitoring plastic pollution and approve a list of disposable goods, containers and packaging that should be phased out in Russia. "It will not be possible to solve the plastic pollution problem any other way," Greenpeace's report concludes [37].

At end of March, Russia was considering giving up disposable plastic tableware at the legislative level, but so far no concrete action has been taken in this direction. On the territory of the Russian Federation, unfortunately, it is very difficult to find a clean sea coast, and the Black Sea coast is no exception. The Azov, Caspian and Black Seas are considered one of the most polluted seas. According to ScanEx company, carrying out research commissioned by the Russian Emergencies Ministry, the Black Sea pollution is quite significant. Environmental situation on the territory of the Black Sea coast requires special attention to the issues related to nature protection and environmental education of the population, as the researchers have established a crisis state of environmental situation: reduction of biological diversity, water pollution by chemicals and many more. "At the same time, the content of plastic in the sand on the Crimean coast is an order of magnitude lower than in the coastal zones of other seas," assured Andrey Bagaev, researcher of the Department of Hydrophysics of the shelf of the Marine Hydrophysical Institute of the Russian Academy of Sciences. RIA Novosti Crimea: [38].

In March 2019, the European Parliament banned the circulation of disposable cutlery, plates, straws, sticks, containers and cups in the European Union from 2021 and set a goal to bring the collection of plastic bottles to 90% by 2029. The example of Europeans will soon be followed by Canada and India. To varying degrees, 127 countries have legislation governing the circulation of disposable goods, containers and packaging (including plastic), according to Greenpeace Russia.

One of the most comprehensive and up-to-date reviews of all microplastics research was compiled for the European Commission by SAPEA experts and scientists in early 2019. It focuses on the analysis of all possible effects of plastic contamination, not just its impact on human health. The Commission's general conclusion was that little is known about the dangers of microplastics today. However, the commission managed to draw several important conclusions:

1. "At least in some places on the planet, the concentration of microplastics exceeds the level at which its impact would be invisible.
2. The number of such places is likely to be small, but it is impossible to quantify them due to lack of data.
3. If the level of plastic accumulation continues to grow at the same rate as it is now, then within a century "risk points" will be most places on the planet".

9.6 Black Sea Pollution with Heavy Metals

Black Sea pollution with heavy metals is not yet significant. However, there are local anomalies near the coast. For example, in areas of viticulture, when treating plantations with copper preparations there may be an excess of toxic copper ions in rainwater afterwards. From the industrial enterprises cadmium, chromium and especially lead, the another source of which are the exhaust gases of automobile transport. Environmental scientists have calculated that with sewage and rainwater flows, about 4 tons of lead, 70 tons of mercury and 10 tons of zinc enter the Black Sea basin annually.

Sea water pollution with heavy metals causes significant environmental damage. Thus, the following heavy metals and toxic chemical elements have been identified as potentially hazardous for humans by the joint UN group of experts on scientific aspects of sea pollution: mercury, cadmium, cobalt, manganese, lead, chromium, zinc, beryllium, vanadium, arsenic, antimony [1].

Experts confirm that there are areas along the Black Sea coast with an excess of toxic copper, cadmium, chromium and lead ions. The cause of heavy metal pollution of the Black Sea is industrial waste water, car exhaust … "The problem is people who have densely populated the Black Sea coast, especially in Russia, Bulgaria, Romania and Ukraine. The Turkish coast is not so densely populated and, most importantly, there are no large rivers flowing into the sea. The problem is also that in Russia and Ukraine, environmental and water protection legislation is not observed. This is not a new problem, it stretches back to Soviet times. There was a small respite in the sea pollution during the collapse of the USSR, when many industrial enterprises reduced their turnover," says the head of the public organization "Clean Coast. Crimea" Vladimir Garnachuk (checko.ru'company/krehoo-chisty-bereg-krym).

Of the above metals, the distribution of mercury in the Black Sea waters has been studied most fully. By toxicological properties, mercury compounds are classified into the following groups: (1) elemental mercury; (2) inorganic compounds of mercury; (3) methyl and ethyl mercury compounds with short chain and other mercury organic compounds. Mercury compounds Hg^0, Hg^+ and Hg^{2+} are interchangeable in the environment.

Mercury in natural waters can be present in three states: elemental (O), univalent (+1) and divalent (+2). The forms of Hg presence and their distribution depend on pH and Eh, as well as on the nature and concentration of anions, which form stable complexes with Hg. In well aerated waters (Eh > 0.5 v), bivalent mercury Hg prevails, while in reducing conditions elementary Hg prevails. The presence of sufficient amounts of sulfides—ions even at very low Eh values stabilizes the bivalent Hg in the form of hydrosulfide or sulfide complexes [1].

The consumption of fish with a high content of methylated forms of mercury in the food was the reason for the so-called Minamata disease. This disease, which claimed the lives of more than 200 people (the total number of victims was several thousand), was first registered in Minamata village (Niigata Prefecture, Japan) in 1953 [3].

According to data [39], mercury concentrations in the Black Sea water were 0.011–0.580 μg/dm^3, and maximum concentrations were observed in water samples taken from the Dnieper-Bug Liman, Dniester and Danube estuaries. Only with the Danube discharge, up to 40–60 tons of mercury is released annually, while with the Dnieper discharge—up to 5 tons of this toxicant. All in all, according to the authors [39], without taking into account the flow from the Turkish territory comes to 66 tons of mercury annually. This half of mercury comes with suspended matter. The authors above also show that self-purification of sea waters from mercury by biosedimentation is not less than 100 tons of mercury per year, with a total content of 14 thousand tons in the sea water column. According to SBSOI, the waters of the Crimean region are the most polluted with mercury—concentrations of more than 0.40 mkg/l, the south-western region along the section of the m. Chersonesos—Bosporus Strait [1].

The content of dissolved mercury in the Novorossiysk water area in 2017 in three analysed samples was below the detection limit (DL = 0.010 μg/dm^3). Iron concentration was determined in seven samples; in one of them it was below the detection limit (DL = 20 μg/dm^3), and in the others it reached 36 μg/dm^3, on average 25.3 μg/dm^3 [19]. The concentration of dissolved mercury in sea water was below the detection limit of the chemical analysis method used (DL = 0.01 μg/dm^3) in all 64 analysed samples in the coastal waters of the Greater Sochi in 2017. In recent years, dissolved mercury in water was only detected in April 2013 with a maximum of 0.0042 μg/dm^3.

Pb. Lead content in coastal waters of the Sochi-Adler region was in the range of 1.1–33.1 μg/dm^3; average annual concentration decreased slightly compared to the previous year to 9.06 μg/dm^3 (in 2016—10.2 μg/dm^3). The maximum value (3.3 MAC) remained at the same level as last year and was recorded in April in the Sochi port surface area. Only 18 samples out of 64 (28%) had lead concentrations above the norm, 13% lower than last year. In general, over the last decade and a half there has been an increase in both average and extreme lead levels in the waters of the district. In 2015–2016 the average concentration exceeded the MAC, and in 2017 it was 0.9 MAC [19].

Fe. The concentration of iron in coastal waters between the mouths of the Mzymta and Sochi rivers varied between 3 and 177 μg/dm^3; the average value was 38.6 μg/dm^3. In 18 samples out of 64 (28%) the values exceeded the MAC, mainly due to the survey conducted on March 22 across the entire study water area. The maximum (3.5 MAC) was recorded at the mouth of the Sochi River in the bottom layer on November 15. The average annual concentration of iron in the water area of the port of Sochi was 44.6; in estuarine areas it was 43.2 and seaward 30.5 μg/dm^3. Average values in the surface and bottom water layers were 36.5 and 40.8 μg/dm^3 respectively. In the last 13 years, a four-year period from 2008 to 2011 was recorded for very high maximum iron concentration values (281–869 μg/dm^3), before and after which the extremes were usually in the range of 1–2.5 MAC. Except for this four-year interval no significant inter-annual changes in neither maximum nor average iron concentration values in the waters of the Greater Sochi area have been recorded (Hydrochemical quality of sea waters. Yearbook 2017. 2018).

The content of heavy metals in bottom sediments is an objective indicator of the contamination of the reservoir and the total anthropogenic load on it. By accumulating pollution, which comes into the body of water for a long period of time, bottom sediments are an indicator of the ecological state of the watershed. Bottom sediments, as a complex multi-component system, play an important role in the formation of hydrochemical regime of water masses and the functioning of ecosystems of reservoirs. They actively participate in the intra-water cycle of substances and energy as habitats for numerous groups of living organisms [40]. The redistribution of metals in bottom sediments is accompanied by the formation of stable technogenic anomalies corresponding to the ecological risk areas for benthic communities and probable secondary contamination of the reservoir [41].

The Sevastopol Bay is classified as an area of high environmental risk, which is due to the restriction of its water exchange with external roads as a result of the construction of protective breakwaters, intensive anthropogenic activities on the coast, wastewater discharge and inflow of contaminants with the river runoff. Black River.

The work [42] identifies elevated concentrations of heavy metals in the bottom sediments of the Sevastopol Bay. The zones of elevated concentrations of such metals as lead and zinc (Fig. 9.7), chromium (Fig. 9.8) are identified in the Southern Bay and in the central part of the Sevastopol Bay, in the areas adjacent to Kilen Bay and Holland Bay. Analysis of the results of the spatial distribution of the

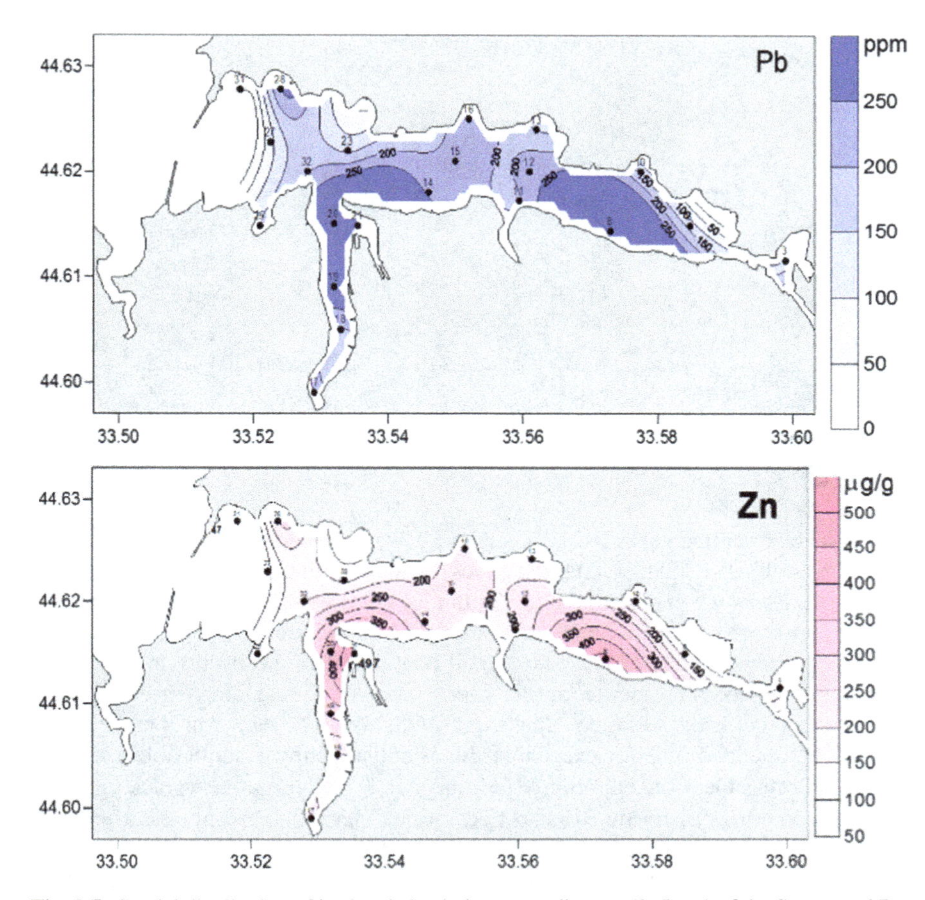

Fig. 9.7 Spatial distribution of lead and zinc in bottom sediments (0–5 cm) of the Sevastopol Bay

studied trace elements shows that the maximum concentrations **As, Cu, Pb, Zn, Cr,** V were found in the Southern Bay and in the central part of the bay, and for some elements—at the top of the bay, where the metal recycling facility is located (assumed source of pollution) [43].

According to research [44, 45], bottom sediments of the Sevastopol bay are represented mainly by fine muddy fraction, sandy siltstones and silted shells. The features of the spatial distribution of the Sevastopol Bay sediments are related to the peculiarities of the flooding character of the Black Sea, which determines the qualitative and spatial heterogeneity of the terrigenous material entering the bay.

According to the data [43] and [45] for the period 2002–2008 the content of lead, cobalt, chromium and iron in the bottom sediments of the water area of the Southern Bay increased. In the area of the central part of the bay the content of chrome and iron in 2016 decreased in comparison with the data obtained earlier [46].

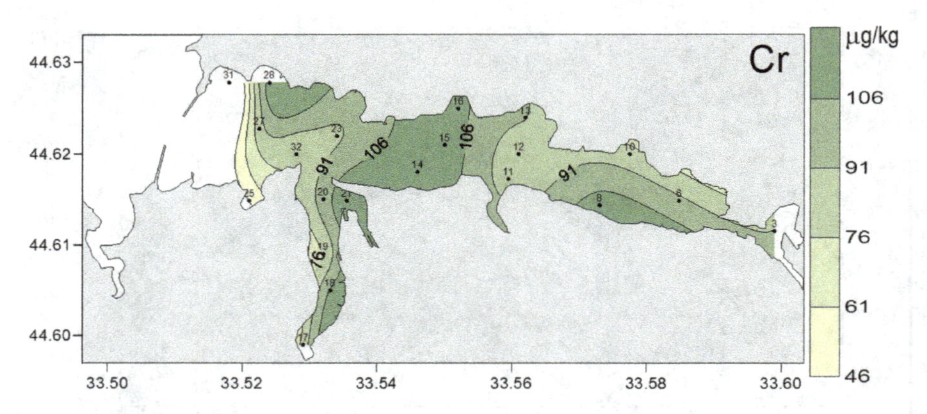

Fig. 9.8 Spatial distribution of chromium in the surface layer of bottom sediments of the Sevastopol Bay

For seabed sediments in Russian territorial waters, there are currently no normatively established quality characteristics of their pollutant concentrations. It is possible to assess the degree of contamination of bottom sediments in the controlled sea area on the basis of compliance of the level of individual pollutants with the criteria of environmental assessment of soil contamination according to regulatory indicators adopted in other countries, for example, the "Dutch sheets" (Neue NiederlandischeListe. Altlasten Spektrum 3/95, Warmer H., van Dokkum R., 2002). The obtained units of exceeding the established upper admissible pollution limits ("Permissible Concentration", PC) are not legal normative values either in European countries or in the Russian Federation. These values only clearly represent the extent to which the actual substance content in the sample exceeds a certain relatively reasonable limit. They can be used to simplify comparative characterization of different water areas or to assess inter-annual variability. In addition to the Dutch Sheets, there are other systems for assessing the quality of bottom sediments around the world.

Radioactive pollution of Black Sea waters. The Black Sea is one of the most radioactively contaminated basins of the World Ocean [47, 48] It ranks second after the Irish Sea in terms of ^{90}Sr content and third in terms of ^{137}Cs pollution, second only to the Irish and Baltic Seas. One of the most dangerous man-made disasters is accidents at reactors or nuclear facilities. Under normal operation of nuclear reactors there are no significant aerosol products entering the atmosphere leading to radioactive fallout. In normal operation, inert gases (^{41}Ar, ^{133}Xe, ^{85}Kr) and a small amount of hydrogen isotopes (^{3}H) and iodine (^{131}I) can be released into the atmosphere. Gaseous products emissions can give (2-4)·10^5 Ci/year, aerosol products—up to 10 Ci/year, radioiodine—5 Ci/year.

Of the known accidents that affected the radioactive contamination of the Black Sea as a result of the release of huge amounts of radionuclides, the world's largest accident at the Chernobyl nuclear power plant (NPP) in 1986 should be attributed to the Chernobyl accident. As a result of a thermal (non-nuclear) explosion at the Chernobyl nuclear power plant on 26 April 1986, reactor unit 4 was destroyed and there was a short-term release of radionuclides. Then, during two weeks (until May 9 inclusive), due to high temperature, radioactive gaseous aerosol products were released in the form of a powerful jet stream into the atmosphere, in which radionuclides ^{239}Np, ^{99}Mo, ^{132}Te, ^{131}I, ^{140}Ba, ^{140}La, ^{141}Ce, ^{103}Ru, ^{95}Zr, 95Nb, ^{144}Ce, ^{106}Ru, ^{134}Cs, ^{89}Sr, ^{90}Sr, ^{91}Y were identified [1].

Shortly after the Chernobyl catastrophe—May 10, 1986 in the zone 10–30 km around the nuclear power plant, measurements of radioactive contamination showed that the largest contributors were isotopes ^{131}I (38% of all radioactivity), ^{103}Ru (14%), ^{95}Zr + ^{95}Nb (22%), ^{141}Ce(7.8%), ^{144}Ce (6.5%), ^{137}Cs (4.7%), ^{106}Ru (3.7%), ^{90}Sr(0.8%). Over time, short lived radioisotopes (^{131}I, ^{103}Ru, ^{95}Zr, etc.) decayed, and 2 years after the Chernobyl accident >90% were radionuclides with a half-life of 27.7 years. The consequences of the Chernobyl accident are still evident in elevated ^{137}Caesium and ^{90}Strontium levels in the areas affected by the fallout [1]. As a result of primary product deposition at the Chernobyl NPP, the ^{137}Cs reserve in the 0–50 m layer of the Black Sea was exceeded 6-10 times as compared to the preaccident one. The inflow of ^{137}Cs into the Black Sea water area in 1986 was estimated at 1.7–2.4 PBq, which meant an increase in its stockpile in the whole volume of the sea by at least 2 times, with extremely heterogeneous contamination of the water area. The northern part of the Black Sea and the coastal areas of the Crimea and Caucasus were the most contaminated. The process of vertical redistribution of ^{137}Cs occurred mainly in the 0–200 m layer, i.e. the ^{137}Cs penetration rate below the main pycnocline was comparable to the rate of its radioactive decay. In the case of multiple entry of ^{137}Cs in the above quantities to the surface of the Black Sea, the time of return of the field of concentration in the active layer to its original state is estimated to be 15–20 years.

After the Chernobyl accident, it was possible to study the migration patterns of long lived ^{137}Cs and ^{90}Sr radionuclides in the Black Sea basin and their dynamics in order to obtain a balance of these radionuclides. On the other hand, after the Chernobyl accident it was possible to study water exchange processes in the sea using the marine conservative $^{134,137}Cs$ and ^{90}Sr as tracers to assess the transport of transformed river water on the north-western shelf and the intensity of vertical water exchange in the deep sea. These approaches have made it possible to determine the role of the Danube and the Dnieper in the pollution of various areas in the north-western part of the sea, to assess the rate of pollution of deep water masses and the capacity of the surface waters to self-clean through the impact of all water exchange processs. In paper [49] presented the results of observations and mathematical modeling showing a significant difference in ^{137}Cs and ^{90}Sr behavior in the Black Sea after the Chernobyl accident, which was that a decrease in ^{137}Cs concentration in the 0–50 m layer 2 times every 5–6 years and in the 50–100 m layer under conditions of insignificant inflow with river runoff was controlled by three processes: vertical water exchange, discharge through the Bosporus Strait and radioactive decay [1].

On the basis of the data of radio-ecological monitoring and nuclear-geochronological reconstruction in the work [50] the dynamics of the Black Sea radioactive contamination for the period 1986–2013 is traced, taking into account the contribution of secondary sources of Chernobyl radionuclides, which include their inflow from watersheds and remobilization from bottom sediments.

It has been established that the totality of these processes, as well as the variability in river runoff, which increased significantly in 1995–1999, could be the main reasons for the increase in the content of ^{90}Sr and ^{137}Cs in water, bottom sediments and hydrobionts of the Black Sea observed in late 1990—early 2000s. The data obtained allowed for substantial refinement of the forecast of the period of time when radioactive contamination of the Black Sea will reach preaccident levels. The results of the determination of ^{90}Sr and ^{137}Cs activity concentrations in the surface water layer of the Black Sea in 2011–2013 showed that the levels of radionuclides in the Black Sea are still relatively high and in some cases exceed the values recorded during the preaccident period, when they were on average about 16 Bq m^{-3} for ^{137}Cs and 22 Bq m^{-3} for ^{90}Sr [50]. The highest concentration of ^{90}Sr was observed in 2011 near the Dnieper-Bugskiy estuary, which indicates the continued flow of this radionuclide into the Black Sea with the discharge of the river. The Dnieper River is the largest concentration of ^{90}Sr in 2011 near the Dnieper-Bugsky Liman, indicating continued inflow of this radionuclide into the Black Sea with the discharge of the Dnieper River. However, unexpectedly high concentrations of ^{90}Sr and ^{137}Cs were also found near the Kerch Strait, which the authors attribute to the influence of Dnieper waters flowing into Crimea from the waters of the North Crimean Canal.

Thus, the combination of processes of delayed inflow of Chernobyl radionuclides with solid river runoff, which increased significantly between 1995 and 1999, is the most important factor in the development of the Chernobyl region, This had a significant impact on the dynamics of the content of Chernobyl radionuclides in the Black Sea and should be taken into account when estimating the time period when it will reach precedent levels.

In aquatic ecosystems, the level of radioactive and chemical substances in water determines the development of organisms in different biotopes, at different levels. This is why scientists so closely monitored the state of the sea, studied how the plutonium concentration changed between 1986 and 2014, and finally concluded that sea waters, especially those close to the surface, have good natural ability to self-clean from this element. Today, the Black Sea is almost pure, with over 90% of all radioactive metal resting on the bottom because of its gravity, and about 9% still dissolved in both surface and near-bottom waters.

In 1986, the plutonium concentration at the sea surface was 12 microbecqueres per cubic meter, in 1988 6.6, in 1998 4.0, and in 2013 only 0.3 microbecqueres, which is close to the background norm. Less than 1% of sea life contains plutonium. The highest concentration is among bottom creatures, but they are less vulnerable to radiation than surface water inhabitants.

In work [51] the comparative analysis of influence of a level of trophicity of sea waters on processes of redistribution of radionuclides of plutonium in sea ecosystems of different scales is carried out: in semi-enclosed coastal water area and in the western part of Black Sea, in open areas of Black Sea and Mediterranean seas. Quantitative characteristics of the radio-ecological processes of sediment deposition of $^{239+240}Pu$ in bottom sediments in the coastal and open areas of the Black Sea and the half-lowering period of $^{239+240}Pu$ in surface Black Sea water were determined. An increase in water trophicity, causing an increase in suspended sediment and deposition rate, has been shown to play a significant role in the redistribution of $^{239+240}Pu$. There has been an increase in the pedotropic properties of plutonium and an increase in the rate of self-purification of Black Sea surface waters from plutonium radionuclides through their transfer to bottom sediments. According to the author [51] the level of trophicity of sea waters influenced the values of radio-ecological parameters of $^{239+240}Pu$ migration in the studied water areas, the radio-ecological status of the sea with respect to plutonium radionuclides and the biogeochemical type of Pu (проверить) behavior in water bodies.

The paper [52] presents the results of the development of criteria for the normalization of anthropogenic impact on the water area of the Sevastopol Bay by the flows of the post-Chernobyl (^{90}Sr, ^{137}Cs, $^{239,240}Pu$) and natural (^{210}Po) radionuclides deposited in the thickness of the bottom sediments under modern conditions of natural biogeochemical cycles. Differentiation of the bay water area to areas with different biogeochemical conditions, as well as the application of a balance approach in the interpretation of materials from natural observations, allowed us to assess the conditioning capacity of the Sevastopol Bay ecosystem. It was assessed by pollutant elimination flows—conserved radioactive substances—into the water depot, which is the open part of the Black Sea, and into the geological depot—the thickness of its bottom sediments.

The main mechanisms for self-purification of the marine environment from conservative radioactive substances are related to the reduction of their concentrations in water through their migration to adjacent waters and sedimentation elimination to bottom sediments. For conservative radioactive contamination, the water mass can also be considered as a depot, whose pool decreases annually by the amount of radioactive decay of radionuclides. In the surface layer of the bottom sediments of the Sevastopol Bay distribution fields of ^{90}Sr, ^{137}Cs, $^{239,240}Pu$, ^{210}Po [52] indicate that the maximum concentrations of pollutants in the bottom sediments are confined to the places of waste water discharge, which confirms the predominant role of sedimentation processes in self-purification of water. Criteria for rationing the anthropogenic impact on the elimination flows from the aquatic environment to the bottom sediments (geological depots) of post-Chernobyl (^{90}Sr, ^{137}Cs, $^{239,240}Pu$) and natural (^{210}Po) radionuclides have been substantiated. According to the authors [52], these criteria can be used in the system of measures to implement sustainable water area development. Determination of water self-cleaning flows with equal maximum allowable concentration (MAC) of contaminants in the ecosystem components makes it possible to estimate the maximum assimilation (or ecological) capacity of water areas with respect to the studied spectrum of contaminants.

9.7 Methane Gas Jets in the Black Sea and Their Impact on the Ecological Status of the Black Sea Ecosystems

Methane is an important element of the Black Sea ecosystem, of natural origin, which can have an impact on global climate change to some extent. Researchers' attention to methane geochemistry in the Black Sea has now increased significantly due to the following circumstances. First, it is already known [53] that the methane concentration in the atmosphere increases by 1% annually, and this cannot be ignored when studying global warming on the planet (greenhouse effect). Although methane is the second greenhouse gas after CO_2 in terms of contribution, it absorbs thermal energy 21 times more effectively than CO_2. Secondly, in the 80s–90s there were publications related to the behavior of methane in the Black Sea. This is the discovery in 1989 [54] of gas emissions on the north-western shelf of the Black Sea, subsequent studies of which [55] showed that the released gas 94–98% consists of methane. Similar gas releases were found in the eastern part of the sea on the coast of Georgia in the area of Poti-Batumi [56] As on the northwestern shelf, the released gas 98% consisted of methane [57].

To date, the main processes of methane biogeochemical behavior in the sea have already been assessed. It is believed that the formation of methane hydrates at the bottom of the deep-sea part of the Black Sea and jet gas emission at the periphery of the basin are components of one global process—the gas emission of the Black Sea bottom.

Observations have identified frequent cases of direct gas flares reaching the sea surface because the solubility of methane in water is very low (1 volume of methane dissolves in 100 volumes of water). The maximum observed height of the flare is 280 m. In total 136 gas manifestations were registered. Comparison of the planned location of gas sources with the results of geological study of the bottom relief indicates that they are confined to the tectonic zones of ruptures, estuaries, shelf edge, continental slope, as well as areas of mud volcanism [57].

In terms of frequency of jet gassing, the Dnieper paleo-irus is one of the most active areas in the Black Sea. Later on, the Dnieper paleodelta is one of the most active areas in the Black Sea. The Dnieper served as a testing ground for numerous studies related to the phenomenon of jet gas emissions, among which it should be noted: analysis of localization, spatial distribution and ecological role of jet gas outputs [58].

Since there are mainly aerobic conditions in shallow water, methane oxidation is carried out by oxygen to the cellular level, passing into the biomass of bacteria, and to carbon dioxide. Thus, it serves as a potential source for primary and secondary organic matter production. Thus, under the conditions of the north-western shelf of the Black Sea, when methane gas emissions intensify during the winter and autumn seasons, its oxidation by oxygen will be more intensive, which, combined with higher gas solubility at lower temperatures, will ensure a minimum flow of methane into the atmosphere. An important factor in methane oxidation is the dissolved oxygen content in water. Maximum rates of methane oxidation are observed in the microaerophilic zone in the water layer with concentrations of 0.1–1 mg/l.

The rate of methane oxidation decreases greatly both when O_2 concentration increases and when it disappears. The temperature regime of life activity of methane oxidizing microorganisms is in the range from 4 to 40 °C.

A completely different situation will take place in the shallow waters of the northwestern part of the Black Sea during the summer season. During the summer season, the north-western Black Sea shelf is at high risk of benthic hypoxia and frosts in the bottom flora and fauna. The occurrence of anaerobic conditions in the north-western Black Sea shelf has a significant impact on methane utilisation processes. Modelling of the hypoxia process has shown that during hypoxia the whole range of reduced compounds is formed in the bottom layer: hydrogen sulphide, sulphur, thiosulphates, ammonium ions, methane, etc. This "additional" methane as a result of vital activity of the archae-sulfate-reduction consortium can become a reserve source of hydrogen sulfide on reaction:

$$CH_4 + SO_4^{2-} \rightarrow HS^- + HCO_3^- + H_2O,$$

which could lead to environmental degradation in the hypoxia zone.

If there are no frosts and the process of methane gas emission increases during the warm period of the year in shallow water conditions, the flow of methane into the atmosphere will sharply increase as a result of reduced solubility of the gas with increasing temperature and slower oxygen oxidation of methane at much lower concentrations in water during the warm period.

Figure 9.9 shows a map of the distribution of main methane sources in the Black Sea. We can see that the jet gasses and cold seeps on the northwestern shelf are located in the zone of possible hypoxia occurrence. In addition, it is known that the methane source is also Danube waters.

In the process of methane release from bottom sediments, microbial communities may consume methane in an anaerobic environment, which significantly reduces its release to the atmosphere, which is biogeochemical since methane is a greenhouse gas. During the anaerobic oxidation of methane in marine sediments, the saturation of sediments with organic matter is important. In sediments with a high organic content, the supply of oxidisers to the water-bottom sediment boundary is generally reduced. Oxidisers (O_2, NO_3^-, Fe(III), Mn(1V) and SO_4^{2-}) penetrate the water-bottom sediments at the water-bottom sediment interface and are used by micro-organisms to oxidise organic matter in a certain thermodynamic order, whereby SO_4^{2-} is used last. When all these oxidizing agents are exhausted, CO_2 becomes the main oxidizer and the oxidation of organic matter is accompanied by methane production. Under these conditions, the depth (zone) in the sediments, where sulfate production is replaced by methanogenesis, is called methane sulfate transition.

This depth corresponds to the methane anaerobic oxidation zone (MAO). Methane anaerobic oxidation according to [60] is performed by at least two different groups of archaebacteria, ANME-1 and ANME-2. These archeobacteria are

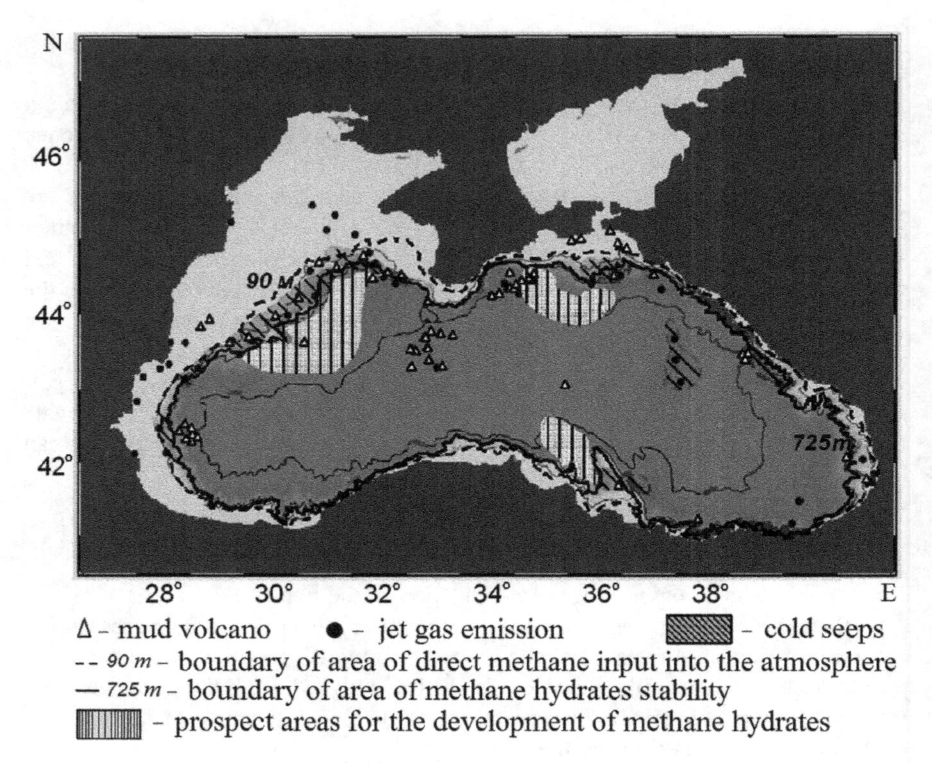

Fig. 9.9 Sources of methane in the Black Sea [59]

often observed in a consortium with sulfate reducing bacteria. The metabolism of this consortium is based on the transportation of hydrogen between communities. There are reliable data on MAO flows not only in the bottom sediments, but also in oxygen-free sulphate waters, such as the Cariaco Depression, Black Sea, etc., where MAO is an important process for reducing methane content in water.

The following work [60] shows the potential reactions that occur in seawater and sediments as a result of archae and sulfate reducing bacteria during methane anaerobic oxidation (MAO).

– for CH_4 consuming archaeobacteria

$$CH_4 + 2H_2O \rightarrow CO_2 + 4H_2$$
$$CH_4 + 4HCO_3^- + 2H^+ \rightarrow CO_2 + 4HCOOH + 2OH^-$$
$$CH_4 + CO_2 \rightarrow CH_3COOH$$
$$2CH_4 + 2H_2O \rightarrow CH_3COOH + 4H_2;$$

– for sulfatereducent catalyzed reactions

$$SO_4^{2-} + 4HCOOH \rightarrow S^{2-} + 4CO_2 + 4H_2O$$
$$SO_4^{2-} + 4H_2 \rightarrow S^{2-} + 4H_2O$$
$$SO_4^{2-} + CH_3COOH \rightarrow 2HCO_3^{-} + H_2S$$

The analyzed conditions of the environment in which the process of methane oxidation took place, showed that the aerobic oxidation of methane, in contrast to MAO, dissolution of carbonate occurs, because CO_2 is a weak acid, with MAO on the contrary, the deposition of biogenic carbonates occurs. These processes can be presented in the form of equations

$$CH_4 + 2O_2 \rightarrow CO_2 + 2H_2O \qquad Alk = 0$$
$$CH_4 + SO_4^{2-} \rightarrow CO_3^{2-} + H_2S + H_2O \quad Alk = 0$$

In the first equation the alkalinity decreases and in the second one the alkalinity of water increases.

Thus, if methane yields are observed in shallow water under oxidized sediments, there may be dissolution of carbonates and reduction of alkalinity of sea water [57].

The depth-integrated production rates and methane oxidation in water and sediments allow estimating methane flows from bottom sediments to water and from water to atmosphere. The work [61] provides quantitative estimates of possible methane inputs to the atmosphere as greenhouse gas for stations located on the shelf and continental slope in the Black Sea. The results of calculations for these stations showed that shallow stations are characterized by higher methane generation rates and low oxidation in bottom sediments, which leads to a significant flow of methane from sediments into the water. A small fraction of methane is also oxidized in the water. As a result, the methane flow into the atmosphere is quite high (for the shelf (27 m deep)—92.5 μMol CH_4/m^2 per day). At the continental slope station, methane "filtration" is more efficient due to oxidation in the water column. As a result, methane flow into the atmosphere is more than halved for the continental slope (120 m deep)—19 μMol CH_4/m^2 daily. The maximum methane flow from water to the atmosphere is typical for the summer season and is 6.8–320 μMol CH_4/m^2.

If methane yields take place in anaerobic conditions of bottom sediments in the presence of sulfates, the MAO process will contribute to the deposition of biogenic carbonates with an increase in alkalinity of sea water.

Methane transport mechanisms in shallow water conditions depending on the source. Remote methods of hydroacoustic diagnostics allow to determine parameters of gas emissions, free gas concentration in water and power of its sources from the shipboard. Their application requires knowledge of sound scattering mechanisms in gas flares, information on acoustic characteristics of gas emissions and background distributions of sound scattering parameters in investigated areas. The main tasks of acoustic diagnostics are: determination of volumes and dispersion of gas emissions, concentration of free gas in water, methane flows from bottom sediments into water column and from water column into atmosphere [63].

As a result of the conducted research, more than 160 sources were discovered, the gas emissions of which are single flares, groups of flares and weak scattered gas emissions [57] (Fig. 9.10).

The analysis of gas flare echograms allows estimating the ascending rate of discrete gas emissions. Lifting speed values vary in the range from 0.04 to 0.34 m/s. The histogram has a two-modal form. About 8–10% of observed emissions have a climbing speed from 0.04 to 0.1 m/s. Free-floating gas bubbles with a diameter of 0.3–0.75 mm correspond to such speeds. In the overwhelming majority of cases (up to 85% of the investigated sample volume) the bubble lifting speed is 0.2–0.3 m/s [64].

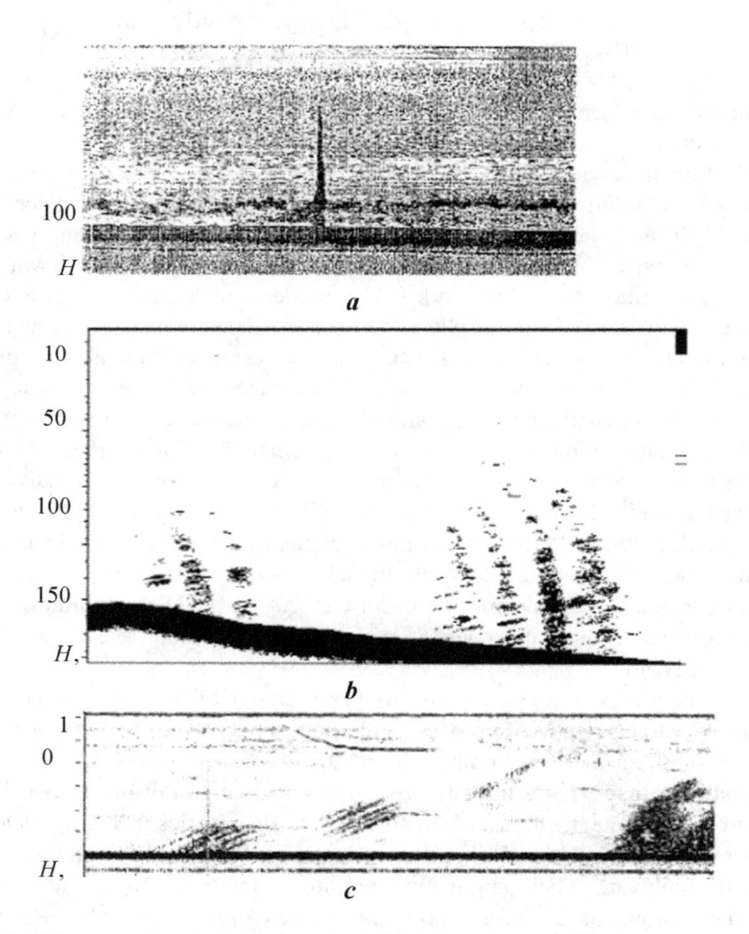

Fig. 9.10 Echogram of a single torch at the inflection of the continental slope (**a**), groups of torches in the paleodelt of the Dnieper (**b**), discrete gas emissions and weakscattered gas in the offshore zone (**c**)

The flow of methane carried into the atmosphere by bubbles is significantly dependent on the initial size distribution of the bubbles. Thus, with finely dispersed emission of methane ($dn \leq 3$ mm) from a depth of 120 m practically all the gas dissolves in water. Calculations have shown that approximately 27% of the initial mass of methane entering water from bottom sources located at the edge of the shelf is carried into the atmosphere by bubbles.

Thus, in the Black Sea the fields of jet gas discharge are located mainly on the shelf edge and continental slope of the Black Sea, in the paleochannels of the Dnieper, Danube, and Don rivers and in the estuary areas of several Caucasian rivers, and can be considered as a source of gas supply to the atmosphere [59].

Higher methane concentrations on the surface occur in the mixing zone of Danube and sea water, as well as in shallow water in areas of sip and jet gas activation. With increasing depth, methane is increasingly effectively "filtered" through oxidation in aerobic and anaerobic conditions. Thus, even deep-water mud volcanic eruptions do not make a significant contribution to methane flow from the water column to the atmosphere. The rates of methane generation and methane oxidation have significant seasonal and inter-annual variability that requires further study.

In the Black Sea gas hydrates, taking into account their stability limit (725 m), methane reserves are estimated at 80 billion m^3 to 49 trillion m^3 [62]. Anaerobic oxidation of methane is of particular environmental importance in the Black Sea, as it is an effective mechanism to prevent methane from entering the water column into the atmosphere.

9.8 International Conventions and Agreements of the Black Sea Countries Aimed at Reducing Sea Pollution

The main international document regulating the protection of the Black Sea is the Convention on the Protection of the Black Sea from Pollution, signed by six Black Sea countries—Bulgaria, Georgia, Russia, Romania, Turkey and Ukraine in 1992 in Bucharest (Bucharest Convention). **Bucharest Convention (Convention for the Protection of the Black Sea from Pollution)**—Convention signed in 1992 in Bucharest (Romania) by representatives of Bulgaria, Georgia, Russia (ratified the Convention on August 12, 1993 by the Resolution of the Supreme Council of the Russian Federation No. 5614-1 "On ratification of the Convention for the Protection of the Black Sea from Pollution", December 2, 1993 the Government of the Russian Federation issued the Resolution No. 1254 "On measures to organize the implementation of the Convention for the Protection of the Black Sea from Pollution"), Turkey, Romania and Ukraine. The Bucharest Convention entered into force on 15 January 1994. The Treaty defines the obligations of these States to reduce and control pollution of the Black Sea and to conduct monitoring and evaluation to protect the marine environment. Specific measures are contained in three protocols that are an integral part of the Convention: the Protocol for the Protection of the Black Sea Marine Environment from Pollution from Coastal Sources; the Protocol on Cooperation in Combating Pollution of the Black Sea Marine Environment by Oil and Other Noxious Substances in Emergency Situations; and the Protocol for the Protection of the Black Sea Marine Environment from Pollution by Dumping.

In 2002, the parties to the Convention signed the Protocol on the Conservation of Black Sea Biodiversity and Landscapes, which includes the "List of Species Important to the Black Sea" [65], Also in June 1994, representatives of Austria, Bulgaria, Croatia, Czech Republic, Germany, Hungary, Moldova, Romania, Slovakia, Slovenia, Ukraine and the European Union signed the Convention on Cooperation for the Protection and Sustainable Development of the Danube River in Sofia. As a result of these agreements, the Black Sea Commission (Istanbul) and the International Commission for the Protection of the Danube River (Vienna) were established. These bodies are responsible for coordinating the environmental programmes implemented under the conventions. Every year on 31 October, the International Black Sea Day is celebrated in all countries of the Black Sea region. The Convention provides for cooperation in scientific research aimed at developing ways and means of assessing the nature and extent of pollution and its impact on the ecological system in the water column and sediments, identifying contaminated areas, studying and assessing hazard factors and developing measures to eliminate them, in particular, alternative methods of treatment, disposal, elimination or use of harmful substances. The participants also agree on the development of monitoring programmes covering all sources of pollution and on the establishment of a pollution monitoring system for the Black Sea (article XV).

On 10 February 2009, in Brussels (Belgium), a working meeting was held between representatives of the Ministry of Natural Resources of Russia (Vladimir Ivlev) and representatives of the Directorate General for Environment of the Commission of the European Communities (Soledad Blanco) to discuss the preparation of a new strategic agreement between Russia and the EU, interaction between the parties under international conventions and organizations, in particular, the Bucharest Convention.

Representatives of the Ministry of Natural Resources of Russia (Vladimir Ivlev) took part in the IV meeting of the Working Group on Water and Marine Issues within the framework of the EU-Russia Dialogue on Environment in Brussels (Belgium), where the EU initiative on accession to the Convention was considered. Following the meeting, the Parties agreed to hold in Moscow in the fourth quarter of 2012 a meeting of EU experts, the secretariats of the Convention for the Protection of the Marine Environment of the Baltic Sea Region and the Bucharest Convention, as well as interested Russian agencies to discuss all aspects of EU accession to the Bucharest Convention.

In order to achieve the objectives of the Convention, the Parties shall establish the Commission for the Protection of the Marine Environment of the Black Sea from Pollution (the Black Sea Commission), which shall be composed of representatives of all States Parties and shall meet at least once a year, and a Secretariat supporting the Commission. The Convention defines the functions of the Commission and the procedure for convening meetings of the Parties (Articles XVII-XIX).

Black Sea Commission (Commission for the Protection of the Black Sea Marine Environment from Pollution)—Commission for the Protection of the Black Sea from Pollution. Together with its Permanent Secretariat, it is an intergovernmental body for the implementation of the Convention on the Protection of the Black Sea against Pollution (Bucharest Convention), its Protocols and the Strategic Action Plan for the restoration and protection of the Black Sea. Expert and information support to the Commission and its Permanent Secretariat is provided by advisory groups on the following activities:

- environmental safety issues in shipping;
- pollution monitoring and assessment;
- to control pollution from land-based sources;
- development of a unified methodology for integrated coastal zone management;
- biodiversity conservation;
- environmental aspects of regulating fisheries and other marine bioresources.

Under the guidance of the Commission, the seven Black Sea regional activity centers are organized on the basis of relevant national institutions. The chairmanship of the Commission is held on a rotational basis in the alphabetical order.

In 1996, the "Strategic Action Plan for the Protection and Restoration of the Black Sea" was signed at the Ministerial Conference on Environment in Istanbul on 31 October, following a comprehensive assessment of the Black Sea. Research has shown that the viability of the sea has been radically depleted over the past three decades. The plan forms a comprehensive guide to practical work to make real improvements to the dramatic state of the Black Sea. At that time, many genuinely believed that the Strategic Plan adopted at the interstate level would indeed help to prevent pollution of the Black Sea and save its inhabitants. But its implementation cannot be called effective. In recent decades, as scientists and specialists from Russia, Turkey, Bulgaria, Romania, Georgia and Ukraine have stated, the anthropogenic load on marine ecosystems has increased so much that in some parts of the coastal area the ability to clean themselves has already been lost. This fully applies to the Russian coastline as well. An example of a purely anthropogenic problem is marine pollution with chemicals. Its main source is rivers. Rivers from 17 countries flow into the Black Sea, the catchment area comes from the good half of Europe and Asian Turkey, and the total annual flow rate of these watercourses is approaching 500 km^3.

On October 31, 2019, the 23rd Black Sea Day took place. Surprisingly, even the inhabitants of coastal countries and cities do not know that there is such a "holiday" at all. In fact, it is difficult to call it a holiday, because the only reason to rejoice is the fact that there were people who were not indifferent, who on this day in 1996 all gathered together in Istanbul and signed the document on the strategic plan of action for saving the Black Sea. Government representatives of the six coastal countries, all but Abkhazia, already knew that our sea was in danger.

The problems of the Black Sea are solved in Russia at the state level. This requires a number of measures aimed at improving the environmental situation as well as significant financial costs. Economic problems are also closely linked to the environment.

- It is necessary to develop a fundamentally new concept of nature management, to create a structure responsible for the Black Sea environmental situation.
- Strict control over the use of trawling and the transition to other ways of catching. Build under water "lying policemen"—massive artificial reefs made of special concrete and without reinforcement inside.
- Tighter control over harmful emissions, commissioning of deep-water sewage collectors.
- Creating living conditions for algae, shrimp, molluscs, which are themselves powerful treatment plants. Construction of underwater habitats.
- Purchase of equipment for coastal strip cleaning.
- Restoration of protective forest belts along the perimeter of agricultural lands and reconstruction of irrigation systems to reduce fertilizer emissions from fields.
- Creation of a modern system for MSW removal and disposal.
- Invention of methods for calculation of material damage caused to the region as a result of misuse of relict forests and coast for construction of oil storage facilities and oil pipelines.

9.9 Conclusions

Uniqueness of the Black Sea is connected not only with its semi-isolation from the World Ocean, but also with a considerable volume of fresh runoff into it, 70% of which comes from the industrial areas of Europe, bringing with it a considerable quantity of various polluting substances, and, at last, with the presence of a hydrogen sulphide zone occupying 87% of the sea volume. The above uniqueness of the Black Sea and the specificity of its water circulation determine mainly the processes of transport, transformation and disposal of pollutants entering the sea area from various sources.

This section analyzes current data on sea pollution levels of oil products, heavy metals, Municipal solid waste (MSW) and plastics, sea radioactive pollution, as well as estimates of nutrient inputs into the sea, as a source of eutrophication of shallow coastal waters, and the spread of this phenomenon in the deep sea basin.

It is shown that with relatively small volumes of emergency spills in the Black Sea, the threat of sea pollution by oil products significantly increases with the construction of new oil storage facilities, as well as the discharge of oil-containing (ballast) water from tankers. The contribution of Black Sea countries to oil pollution of the Black Sea ecosystem is analyzed. At the same time, zones with increased dynamics of OH content are confined to coastal areas and areas of transport highways, which, depending on the season, occupy larger or smaller areas. Estimates of oil contamination of the water area of the South coast of Crimea, the Sevastopol Bay and the Eastern Black Sea coast are given. Information on the Black Sea Track Web (BSTW) operational system for forecasting the spread of oil spills in the Black Sea is given, and estimates of its practical implementation are made.

The level of shallow sea eutrophication was estimated using the example of the north-western shelf of the Black Sea using modern methods of mathematical modeling, which made it possible to determine the causes and consequences of formation of hypoxia and suffocation zones in the studied water area. The role of the Danube River waters is assessed, which, as a result of the capture of the main Black Sea current (MBSc), cause pollution and eutrophication of its western branch, which causes enormous losses to the recreational complexes of Romania and Bulgaria.

For the South Coast of Crimea and the East Coast of the Sea, modern data on nutrient element flows and the main coastal sources of nutrient inputs are presented. It is shown that the eutrophication level is related to local releases of domestic untreated or poorly treated wastewater into the coastal waters of Black Sea towns. The most heavily polluted bays and bays of large cities are Sevastopol, Yalta, Novorossiysk, Gelendzhik, Sukhumi, Poti, Batumi and Pitsunda.

Modern estimates of the level of beach and coastal pollution by solid waste (MSW) and plastics are given. The level of microplastics contamination of beaches in Sevastopol and the East coast of the sea is analyzed. The results of the audit of plastic wastes and their morphology carried out by Greenpeace in Russia in November 2018 on the Black and Azov Seas are considered.

It is shown that in general, a significant part of the Russian Black Sea and Azov Sea coasts today have an index of "unfavourable waters", and this problem should be addressed on a national scale. Of course, the future of coastal ecosystems and the shelf depends not only on the six Black Sea countries, but also on what environmental policy will be implemented by their immediate neighbours.

The section also analyses the level of heavy metal pollution in the sea, which is not yet significant for the sea as a whole. However, there are local anomalies near the coast that can cause significant environmental damage. Information is provided on pollution of the coastal waters of the north-western shelf, the waters of the South Coast of Crimea and the eastern coast of the sea with mercury, lead and iron. The level of contamination of shallow sea bottom sediments with heavy metals as a possible source of secondary contamination of the water body is also analysed. The spatial distribution of heavy metals in the Sevastopol Bay bottom sediments is considered in detail depending on their fractional composition and the location of possible sources of heavy metals.

The radioactive contamination of the sea is considered in the context of assessments of the consequences for the sea of the world's largest accident at the Chernobyl nuclear power plant (NPP) in 1986, which so far have been manifested in elevated levels of ^{137}Caesium and ^{90}Strontium in areas exposed to radioactive fallout. The most contaminated areas were the northern part of the Black Sea, the coastal areas of the Crimea and the Caucasus. The role of the Danube and Dnieper rivers in the contamination of various areas of the north-western part is described, with an assessment of the rate of contamination of deep water masses and the ability of the sea's surface waters to be self-cleaning by all water exchange processes. Estimates of the forecast period of time when the radioactive contamination of the Black Sea will reach preaccident levels are given.

The section also deals with methane as an important element of the Black Sea ecosystem of natural origin, which may to some extent affect the ecological condition of the Black Sea ecosystems in both shallow and deep sea areas, while being a greenhouse gas, influencing global climate change to some extent. It is shown that jet methane gas emissions from the bottom of the Black Sea are confined to estuaries, shelf edges, continental slopes and areas of mud volcanism. Methanogenesis of methane and features of its oxidation processes in aerobic and anaerobic environments are considered in detail. Methane flows from bottom sediments to water and from water to the atmosphere are analyzed quantitatively. The special ecological role of the process of anaerobic methane oxidation, which is an effective mechanism preventing methane from entering the water column into the atmosphere, is emphasized.

The section concludes with a review of international conventions and agreements of the Black Sea countries aimed at reducing sea pollution. The main international document regulating the issues of protection of the Black Sea is the Convention on Protection of the Black Sea from Pollution signed by six Black Sea countries—Bulgaria, Georgia, Russia, Romania, Turkey and Ukraine in 1992 in Bucharest (Bucharest Convention). Information is provided on the content of the three Protocols for the Protection of the Black Sea Marine Environment as an integral part of the Convention. The tasks of the Black Sea Commission (Commission for the Protection of the Black Sea Marine Environment from Pollution)—Commission on Protection of the Black Sea from Pollution—are considered. The Commission, together with its Permanent Secretariat, is an intergovernmental body for the implementation of the Convention on the Protection of the Black Sea against Pollution (Bucharest Convention). It provides information on solving problems of the Black Sea in the Russian Federation on the state level.

References

1. Ivanov VA, Pokazeev KV, Sovga EE (2006) World ocean pollution. Moscow State University, 163 p
2. Vinogradov ME, Shoemaker VV, Shushkina EA (1992) Black sea ecosystem. M. Nauka, 112 p
3. Sovga EE (2005) Pollutants and their properties in the natural environment. Sevastopol, MGI NASU, PF Lomonosov Moscow State University, 237 s
4. https://krym-sea.ru/news/neft-v-krymu.html

5. Solovyova OV, Tikhonova EA, Mironov OA (2017) Content of oil hydrocarbons in coastal waters of the Crimean Peninsula. Scientific Notes of the Vernadsky Crimean Federal University 3(69) (3):147–155
6. blacksea-education.ru'zagr.shtml
7. https://www.oilexp.ru/news/russia/chernoe-more-proektiruyutsya-novye-marshruty-transporta-nefti/7670/
8. lithium.rf's ecology novorossiysk/black sea
9. Kuznetsov AN, Fedorov YA, Zagranichnyi KA (2013) Oil pollution of the Black Sea coast in the Novorossiysk city area (based on the results of the long-term investigations). Izvestia of the higher educational institutions. North-Caucasian region. Natural Sciences. Rostov-on-Don, Southern Federal University, pp 71–77
10. http://www.nmtp.info/holding/press-centre/news/news_detail.php?ID=8577
11. Matishov GG, Stepanian OV, Kharkov VM, Sawyer VG (2014) Modern data on the Azov and Black Seas pollution by oil hydrocarbons (in Russian). Bulletin of the Southern Scientific Centre, pp 49–52
12. Kuznetsov AN, Fedorov YA (2011) Oil pollution in water ecosystems. Regularities of natural transformation. Saarbrucken: LAP Lambert Academic Publishing, 196 p
13. Kuznetsov AN, Fyodorov YA, Zagranichny KA (2011) About the results of the three-year monitoring of the fuel oil spill in the Kerch Strait. North-Caucasian region. Natural Sciences, Rostov-on-Don, Southern Federal University, 4:90–95
14. en.wikipedia.org›International Association of Independent Tanker Owners
15. https://pandia.ru/text/77/280/67433.php
16. Nosenko GN, Butyrskaya IB, Ivanov SV (2015) Modern problems of black sea pollution by oil products and participation of microflora in biodegradation processes of oil pollution. In: Collection of scientific proceedings on the results of international scientific and practical conference "Actual Problems and Achievements in Medicine" Samara, April 2015, S.I. Georgievsky Medical Academy KFU - Simferopol, - Innovation Center for Development of Education and Science, pp 48–51
17. The state of the marine environment: regional assessments. UNEP/GPA, 2006
18. Hydrology and Hydrochemistry of the Seas Volume 1U Black Sea, Exhibit 3 Current state of Black Sea water pollution, 1996
19. Korschenko AN (ed) (2018) Quality of sea waters on hydrochemical indicators. Yearbook 2017. Moscow, "Nauka", 220 p
20. Korotaev GK, Ratner YB, Kubryakov AI (2012) National module of the Black Sea forecasts as an element of the European system (in Russian). Science and Innovation T.8(1):5–10
21. Kubryakov AI, Korotaev GK, Dorofeev VL et.al (2012) Black sea coastal forecasting system. Ocean Sci 8:183–196
22. Belokopytov VN, Kubryakov AI, Pryakhina SF (2019) Modelling of water pollution propagation in the Sevastopol Bay. Phys Oceanogr [e-journal] 26(1):3–12. https://doi.org/10.22449/1573-160x-2019-1-3-1210
23. Trukhin VI, Pokazeev KV, Kunitsyn VE (2005) General and ecological geophysics (in Russian). MPEI V.I. Pokazeev, Fizmatlit, 576 p
24. https://www.livelib.ru/author/110102/series-truhin-vi
25. Belyaev VI, Sovga EE, Lyubartseva SP (1997) Modeling of the bottom hypoxia and appearance of the hydrogen sulfide lenses on the north-west shelf of the Black Sea (in Russian). Reports of the NAS of Ukraine 4:117–121
26. Belyaev VI, Sovga EE, Lyubartseva SP (1997) Modelling the hydrogen sulphide zone of the Black Sea. Ecol Model 96:51–59
27. Yunev OA (2020) Secondary eutrophication of the Black Sea shelf. Ecol Saf Coast Shelf Sea Zones 2:80–91
28. Sovga EE, Godin EA, Plastun TV, Mezentseva IV (2014) Long-term variability of the biogenic elements content in the Yalta Bay (in Russian). MGZh 4:48–59

29. Ivanov VA, Sovga EE, Mezentseva IV (2019) Multi-year dynamics of biogenic elements and oxygen in the water area of Yalta port for the period 2013–2017. Ecol Saf Coast Shelf Sea Zones 2:86–93
30. Selifonova JP, Chasovnikov VK (2013) Zoobentos of the port water areas of the north-eastern Black Sea shelf and its connection with the bottom sediments pollution (in Russian). Water Chem Ecol 1:79–86
31. Selifonova JP, Chasovnikov VK, Yakushev EV (2007) Study of the sea shelf ecosystems in the conditions of high anthropogenic pollution on the example of the Novorossiysk port of the Black Sea. In: Proceedings of the international scientific conference "Large marine ecosystems of Russia in the era of global changes (climate, resources, management)" Rostov on Don, 10–13 Oct 2007, pp 257–263
32. nacep.ru'ekologiya/ekologiya-chernogo-morya.html
33. Sibirtsova EN (2018) MICROPLASTIC pollution of the soils of Sevastopol beaches in the summer period 2016–2017. Ecol Saf Coast Shelf Sea Zones (1):C.64–73. https://doi.org/10.22449/2413-5577-2018-1-64-73
34. Barnes DKA, Galgani F, Thompson RC, Barlaz M (2009) Accumulation and fragmentation of plastic in global environments. In: Barnes DKA, Galgani F, Thompson RC, Barlaz M (eds) Philosophical transactions of the royal society of london. Series B, 364(1526), pp 1985–1998
35. Imhof HK, Schmid J, Niessner R, Ivleva NP, Laforsch C (2012) A novel, highly efficient method for the separation and quantification of plastic particles in sediments of aquatic environments. In: Imhof HK, Schmid J, Niessner R, Ivleva NP, Laforsch C (eds) Limnology and oceanology. Methods, vol 10, pp 524–537
36. svoboda. org'a/30037034.html
37. greenpeace.ru'blogs/2019/07/03…na…chjornogo-i…ili
38. RIA Novosti Crimea: https://crimea.ria.ru/society/20181030/1115489344.html
39. Egorov VN, Polikarpov GG, Svetasheva SK (1992) Mercury contamination of the Black Sea with river runoff and ability of its waters to self-clean as a result of biogeochemical cycles. In: International conference assessment of land-based pollution sources of the seas washed by the CIS States, Sevastopol, , pp 61–62
40. Raininin VN, Vinogradova GN (2002) Technogenic pollution of the river ecosystems. Scientific World, Moscow, 140 p
41. Emelyanov VA, Mitropolsky AY, Nasedkin EI, Pasynkov AA, Stepaniak YD, Shnyukova EE (2004) Geoecology of the Black Sea shelf of Ukraine - Kiev ("Akademperiodika"), p. 296
42. Ovsyanyi EI, Romanov AS, Ignatieva OD (2003) Distribution of heavy metals in the surface layer. Marine Ecol J 2(2):85–93
43. Ovsyanyi EI, Kotelianets EA (2011) Peculiarities of distribution of arsenic and heavy metals in thicker than precipitation of sevastopol bay (in Russian). Ecological safety of the coastal and shelf zones and complex use of the shelf resources. Collection of scientific articles of MGI NAS of Ukraine, 2011, vol 22, pp 296–302
44. Orekhova NA, Konovalov SK (2009) Polarography of the Sevastopol Bay bottom sediments. Marine Ecol J 2:52–66
45. Romanov AS, Orekhova NA, Ignatieva OG et al (2007) Influence of the physical and chemical characteristics of the bottom sediments on the microelements distribution on the example of the Sevastopol bays (Black Sea) (in Russian). Ecol Sea 73:85–90
46. Gurov K, Kotelyanets E, Tikhonova E, Kondratev S (2019) Accumulations of trace metals in bottom sediments of the Sevastopol Bay (Black Sea). In: 19th International Multidisciplinary Scientific GeoConference Surveying Geology and Mining Ecology Management, SGEM 2019, Conference Proceedings. ISBN/ISSN: 1314-2704, 30 June–6 July 2019, Albena, Bulgaria, vol 19, issue 3.1, pp 649–656
47. Polikarpov GG, Egorov VN, Gulin SB, Yin (2008) Radio-ecological vidguk of the Black Sea on Chornobyl catastrophe (in Russian). Bull NAS of Ukraine 4:29–43
48. Egorov VN, Gulin SB, Polikarpov GG, Osvath I (2010) Black sea. In: Atwood DA (ed) Radionuclides in the environment. Wiley, Chichester (UK), pp 430–452

49. Egorov VN, Polikarpov GG, Osvas I, Stokozov NA, Gulin SB, Mirzoeva NY (2002) Radioecological response of the Black Sea to the Chernobyl nuclear accident in respect of long lived radionuclides of caesium -137 and strontium -90. Marine Ecol J 1(1):5–15
50. Gulin CB, Mirzoeva NY, Sidorov IG, Proskurnin VY, Gulina LV (2013) Secondary contamination of the Black Sea with technogenic radionuclides after the Chernobyl accident. NAS of Ukraine reports series Ecology, vol 10, pp 184–190
51. Tereshchenko NN (2017) Influence of sea waters trophicity on migration and deposition of technogenic radionuclides of plutonium. J Siberian Fed Univ Biol T.10(1):20–34
52. Egorov BN, Gulin SB, Malakhova LV, Mirzoeva NYu, Popovichev VN, Tereshchenko NN, Lazorenko GE, Plotitsyna OV, Malakhova TV, Proskurnin VYu, Sidorov IG, Stetsyuk AP, Sidorov LV, Gulina V (2018) Water quality normalization of the Sevastopol Bay by deposition flows of the pollutants into the bottom sediments. Water Resour 45(2):188–195. https://doi.org/10.7868/s0321059618020086
53. Reeburch WS, Ward BB, Whalen SC et al (1991) Black Sea methane geochemistry. Deep Sea Res 38(Suppl 2):1189–1210
54. Polikarpov GG, Egorov VN, Nezhdanov AI (1989) Phenomena of the active gas emission from the uplifts on the depths bed of the western part of the Black Sea (in Russian). Dokl. of the Academy of Sciences of the Ukrainian SSR. 1989. B, vol 12, pp 13–15
55. Shnukov EF, Pasynkov AA, Klimenko SA (1998) Gas flares nature on the north-west shelf of the Black Sea (in Russian). In: Geology, geophysics and hydrography of the north-west of the Black Sea. Omgor NAS of Ukraine, Kiev, pp 69–74
56. Tkeseshashvili GI, Egorov VN, Mestvirishvili SA (1997) Methane gas emission from the Black Sea bottom in the river Supsa near the Georgian coast (in Russian). Geochemistry 3:331–335
57. Sovga EE, Lyubartseva SP, Lyubitskiy AA (2008) Investigation of the biogeochemistry and the methane transfer mechanisms in the Black Sea (in Russian). MGFZh 5:C. 40–56
58. Artemov YG, Egorov VN, Polikarpov GG, Gulin SB (2007) Methane emissions into hydro- and atmosphere by jet gas emissions near paleodelt of the Dnieper River in the Black Sea
59. Egorov VN, Polikarpov GG, Gulin SB et al (2003) Modern ideas about the environment-forming and ecological role of the methane gas jets from the Black Sea bottom (in Russian). Marine Ecol J 3:C. 5–26
60. Valentine DL (2002) Biogeochemistry and microbial ecology of methane oxidation in anoxic environments: a review. Antonie van Leeuwenhoek, vol 81, pp 271–282
61. Elena S, Liubartseva S (2016) Sources of methane and seasonality in methane oxidation on the north-western shelf of the Black Sea. In: Geophysical Research Abstracts, vol 18, EGU2016–6761. EGU General Assembly 2016
62. Shnukov EF (2005) Methane gas hydrates in the Black Sea. Geology Ocean Miner 2:C. 41–52
63. Lyubitsky AA (2003) Hydroacoustic research of the phenomena of active gas emission in the north-western part of the Black Sea. Ecological safety of the coastal and shelf zones and integrated use of the shelf resources. ECOSY-Hydrophysics, Sevastopol, Issue 9, pp 226–240
64. Artemov YG (2014) Distribution and flows of the methane jet gas emissions in the Black Sea. Cand Sevastopol, pp 152 c
65. Mokk A, Mihova IV (2006) Legal regulation of international cooperation of the Black Sea states in the field of marine environment protection. Sumi, Sumi, pp 98–104

Chapter 10
Modern Methods for Assessing the Self-cleaning Capacity of Marine Ecosystems in Shallow Waters of the Black Sea: Ports, Bays, Estuaries

10.1 Introduction

The environmental well-being of shallow marine ecosystems, regardless of the environmental measures taken, is primarily determined by their self-cleaning capacity. Assessment of the self-cleaning capacity of shallow water areas can be done by calculating their assimilation capacity (AC) in relation to a priority pollutant (PP) or complex.

The main natural ways of removing pollutants, providing the process of self-cleaning of the ecosystem, are hydrodynamic transport, physico-chemical and biochemical transformation and deposition in bottom sediments. Regardless of the dominance of any of these routes in different areas of the sea, in general, all of them contribute to reducing seawater pollution to the natural state of the ecosystem.

Despite the ability of the marine ecosystem to withstand, to a certain extent, both fluctuations of ordinary natural factors and changes in living conditions under the influence of anthropogenic impact, its ecological state is largely determined by the volume of incoming pollutants. Since the ability of the marine ecosystem to transform the products of technogenesis is not infinite, it is crucial to determine this limit, i.e. the maximum number of Sounds, which will not cause irreversible consequences for the ecosystem.

In this section of the book, self-cleaning ability is assessed for shallow water areas under intensive anthropogenic load, the so-called sea impact zones. Impact zones of the seas and oceans are considered to be the areas characterized by significant contamination of waters and bottom sediments as a result of intensive anthropogenic impact. In the Black Sea these are the areas of ports, bays, estuaries with difficult water exchange with the open sea, as well as shipping routes, oil and gas production areas, etc. These areas are characterized by high concentrations of organic pollutants (OP): oil and/or phenolic compounds, synthetic surface active substances (SSAS), etc. It is important to decrease the supply of nutrients that cause eutrophication of ecosystems. The simultaneous presence in seawater of various

K. Pokazeev et al., *Pollution in the Black Sea*, Springer Oceanography,
https://doi.org/10.1007/978-3-030-61895-7_10

pollutants, including heavy metals, contributes to the formation of synergistic complexes between them and environmental components. This leads to the degradation of the biotic component both within and adjacent to ecosystems, and prevents its rational use for recreational or fisheries purposes.

For the ecological well-being of impact zones against the background of intensive anthropogenic load and difficult water exchange with the open, less polluted part of the sea, the threat is not only high content of pollutants (P), but also the emergence of fields of "chronic" pollution there. The total impact of low intensity factors (doses of resistant long-lasting chemical compounds) has already caused the degradation of a number of coastal areas, the biotic component in both the boundaries of ecosystems and adjacent waters.

10.2 Analysis of Existing Methods for Calculating the Self-cleaning Capacity of Marine Ecosystems

To assess the self-cleaning capacity of ecosystems in impact zones of the seas, this section of the book examines the methodological peculiarities of implementing two methods for quantitative determination of self-cleaning capacity of marine ecosystems (balance and synoptic) to calculate the assimilation capacity of shallow marine ecosystems. In addition to the analysis and evaluation of the existing methods of AC calculation, examples of their implementation are given for the shallow waters of the Black Sea, which differ in the level of anthropogenic load, as well as in hydrologic and hydrochemical and geographical characteristics. The example of separate sea areas shows the expediency of spatial (zoning) and temporal (seasonality) detailed assessment of the assimilation capacity of ecosystems in relation to various pollutant complexes. The necessary conditions for the application of the balance and "synoptic" methods for the calculation of the assimilation capacity are described, indicating the advantages and difficulties of each. The possibilities of diagnosing the ecological well-being of the ecosystems under study, taking into account AC calculated by two methods, are shown.

Current methods for assessing the environmental condition of water areas actually only characterize the level of marine pollution. Quantitative assessment of the ability of the marine ecosystem to self-cleaning, allowing to normalize the inflow of pollutants into the aquatic environment, can be made on the basis of the concept of the ecosystem's assimilation capacity (AC) developed by Israel and Tsyban [1, 2].

The ability of an ecosystem to withstand the addition of some amount of pollutants (P) without any significant biological consequences was first described in terms of the assimilation capacity (AC) of Cern [3]. This problem was later developed in more detail by Israel and Tsyban [2, 4, 5]. According to Yu. A. Israel "assimilation capacity characterizes the ability of the ecosystem to dynamic accumulation of toxic substances, as well as the possibility of their active removal with the preservation of basic properties of this ecosystem". That is, the ability of the ecosystem through a wide range of biological, physical and chemical processes to provide protection from alien interference is its natural "immunity", the measure of which can serve AC. In this definition, AC has the dimension of a substance flow—the mass of a substance in a unit of volume attributed to a unit of time.

The basis of the balance method of AC calculation for marine shallow water ecosystems included zoning of the investigated water areas by the level of anthropogenic load, the algorithm of AC estimation, refined for the sea areas subject to long-term state departmental hydrochemical monitoring. In the absence of long-term state monitoring data the possibility to assess the self-cleaning capacity of shallow water areas using synoptic method based on the results of a specific oceanographic survey is shown.

The AC concept developed by Yu.A. Israel and A.V. Tsiban, based on the results of versatile oceanological studies, has been tested on the Baltic Sea ecosystem for benzene(a) pyrene, polychlorobiphenyls and a number of toxic metals (Cu, Zn, Pb, Cd, Hg) [1]. AC characterizes the ability of a marine ecosystem to withstand the addition of some amount of SNF without the development of irreversible biological effects.

The actual complexity of AC assessment has two aspects. The first relates to the difficulty of determining the causal relationship between pollution and biological effects. The second aspect is related to the difficulties of forward and backward transition from flow units (in which AC is measured) to mass units in which pollutant concentration (PC) in the environment is measured.

In order to overcome the difficulties of the first aspect, the search for the most vulnerable link in the ecosystem is resorted to and the threshold of its vulnerability in general is set [6]. To overcome the difficulties of the second aspect, the balance method is used [1], which is not always possible, since observations of the pollution of most seas have ceased to be regular, and the intervals between observations significantly exceed the time of renewal of the internal environment of the objects under study.

The final formulas for estimating the mean value $\bar{A}_{mi}$ and standard deviation of the $\sqrt{D[A_{mi}]}$ assimilation capacity of the marine ecosystem (*m*) in relation to the *ith* pollutant are as follows:

$$AE_{mi} = \bar{A}_{mi} \pm \sqrt{D[A_{mi}]}, \quad \bar{A}_{mi} = \frac{Q_m \cdot C_{thr\,i}}{C_{\max i}} \cdot \bar{v}_i,$$

$$D[A_{mi}] = \left(\frac{Q_m \cdot C_{thr\,i}}{C_{\max i}}\right)^2 \cdot D[v_i]$$

where Q_m is the volume of water in the calculated area; $C_{thr\,i}$ is the threshold concentration of the pollutant; $C_{\max i}$ is the maximum concentration of the pollutant in the ecosystem; v_i is the rate of removal of the pollutant from the ecosystem, the average value $\bar{v}_i$ and dispersion of $D[v_i]$ which is determined by the original algorithm [7].

AC assessment of the Sevastopol Bay ecosystem was performed using the balance method according to the algorithm [8] presented in Fig. 10.1.

The algorithm used to calculate AC does not require additional targeted field and laboratory studies, and is therefore more accessible for assessing the self-cleaning capacity of a particular marine ecosystem. However, it is limited by a number of conditions. Firstly, sufficient data series of systematic (monitoring) observations of the content of the recommended for the water area under study complex of pollutants (P) are required. Representation of baseline data according to this algorithm can be achieved by using weekly and monthly observations. Depending on the frequency of monitoring studies the time series should contain observations for the period from three to ten years. Secondly, based on the differences in hydrological indicators, location and power of sources of pollutants, it may be appropriate to conduct a zoning of the water area.

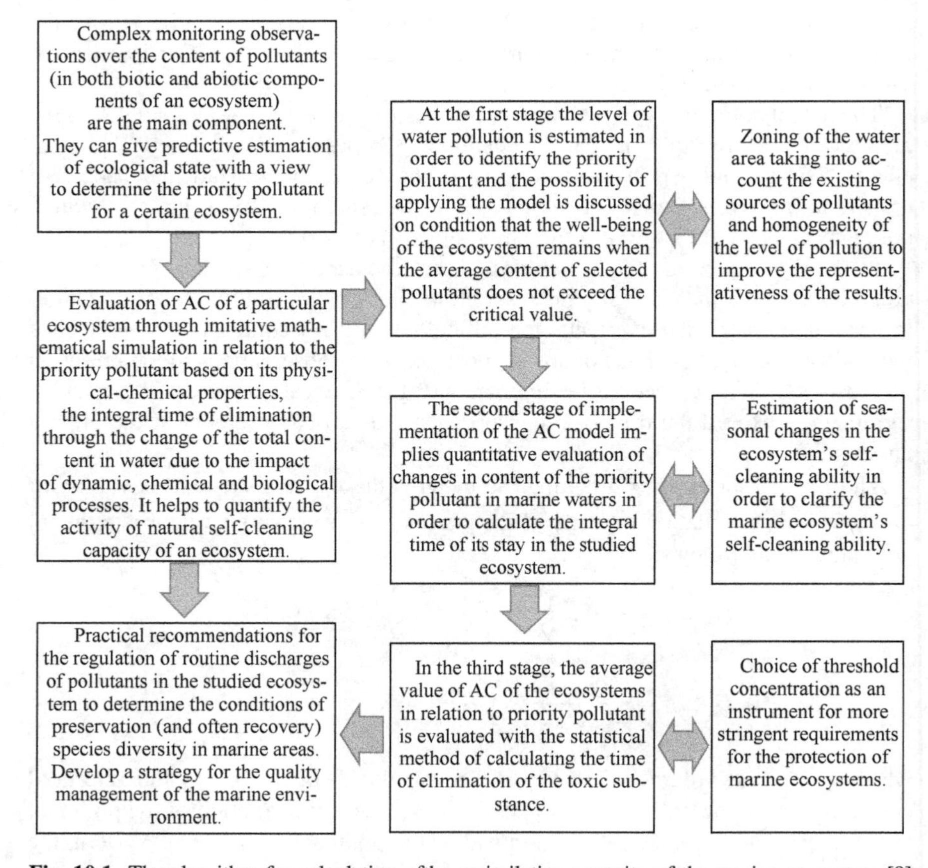

Fig. 10.1 The algorithm for calculating ofthe assimilation capacity of the marine ecosystem [9]

The self-cleaning capacity of bay and bay ecosystems linked to the open sea by narrow straits has been significantly reduced due to limited water exchange in the open sea. Therefore, calculating the dynamics of water area and water exchange through narrow straits or canals is particularly important in normalizing anthropogenic pressures on marine ecosystems of enclosed waters.

For marine areas not covered by state or departmental monitoring observations, AC ecosystems can be estimated using the "synoptic" method [10], for the implementation of which the data from one oceanographic survey are sufficient. The method is based on the assumption that the inhomogeneous distribution of pollutants detected on the basis of survey results in a water mass homogeneous in terms of physical parameters is a consequence of self-cleaning processes taking place in the water mass, the starting point for which is the passage of the last storm in the water area. This method was implemented during the AC assessment of the micro polygon water area in the coastal zone of the Heraclean Peninsula in relation to the petroleum products (PP) according to the oceanographic survey results on the vessel "Turquoise" during two expeditions on May 20–21, 2016 and September 12–13, 2016. Homogeneous water mass contouring on the micro polygon was performed based on salinity and temperature measurements data. The results of AC calculation using this method are given in the paper [11]. This method allowed to calculate the admissible ecological load on the investigated water area and the time required to bring the concentration of the investigated pollutants to Maximum allowable concentration (MAC). The results of the calculation will be discussed in more detail below.

As anthropogenic impacts in coastal ecosystems have already disrupted the natural reproduction and environmental-forming functions of living matter, the problem of rationing the flows of anthropogenic pollution substances has come to the fore when developing a strategy for marine environmental quality management. According to (Egorov V.N. Normalization of the anthropogenic pollution flows in the Black Sea regions by the biogeochemical criteria [12], an important factor in solving the problem of marine AC is the restriction of pollutant flows by biogeochemical criteria reflecting the concentration and sorption functions of living and bone matter, the intensity of mass exchange with adjacent water areas, and the production and sedimentation characteristics of marine ecosystems. When studying AC of the marine ecosystem physical, chemical and biological processes of transformation of a specific pollutant are considered together, which determine the activity of natural self-cleaning.

A characteristic feature of biogeochemical criteria is the regional principle of their applicability. In coastal areas where the natural seasonal amplitude of changes in all processes is quite large, it is particularly difficult to assess AC ecosystems. However, it is possible to approach the solution of this problem by determining the allowable ecological loads in relation to at least one or several SEAs, most typical for the studied area.

10.3 Self-cleaning Capacity Assessments of the Dnieper Estuary and Odessa Port Ecosystems

First of all, coastal sea areas with a high degree of technogenic load were studied, for which rationing of pollutant intake with industrial and domestic discharge per unit of discharge, ignoring their total flow and the ability of the water basin as a whole to self-cleaning, may be inadequate and lead to a significant deterioration of the environmental situation. A positive aspect for such water areas can be considered monitoring observations of water quality within the framework of state and international programs on environmental protection, which allow to trace changes in the content of priority pollutants, to assess the time of their elimination, including in case of need and taking into account seasonal changes.

The waters of the Dnieper estuary and the port of Odessa selected for the study, despite the significant difference in morphometric characteristics, reflect the main physical and chemical conditions for the formation of pollution levels of marine ecosystems inherent to the Black Sea impact zones. Such zones are characterized by extremely high pollution of waters and bottom sediments due to intensive anthropogenic load, including the active work of water transport, discharges of industrial and municipal enterprises located on the coast, storm runoff from urban and agricultural areas, both directly into the water area and indirectly with river waters.

When assessing AC for large water bodies with complex geographic and hydrologic-hydrochemical characteristics, including the presence of fresh water sources and zones of their mixing with sea waters, as well as heterogeneity of anthropogenic load, it is necessary to pre-zoning the water area. In this case AC is calculated for each allocated area. As an example, we consider the Dnieper estuary, the zoning of which is presented in the paper [13], where the determining factor of pollution is the input of pollutants with river runoff and geomorphological features of the water area.

The Dnieper Firth, located in the Northern Black Sea region, is an integral part of the Dnieper-Bugsky estuary. It is united with the Bug estuary by the peculiarities of hydrological regime—distribution of temperature, salinity of estuarine waters, wind wave dynamics [14]. The analysis of the ecological state of waters in the studied water areas is presented in the paper [15], according to which the waters were characterized as polluted or dirty according to the WPI index.

Variation of specific rate of oil products(OP) elimination in different regions of the Dnieper liman according to [15] was determined first of all by intensity of hydrodynamic processes. The average specific value of AC for the entire Dnieper estuary ecosystem with respect to the OP can be estimated at 1.94 mg per year, for the adjacent part of the Black Sea—at 2.74 mg per year. The regionally differentiated approach to the assessment of the self-cleaning capacity of the estuary revealed that for the strait of the western region, which is locked in the narrowness, the AC value was 1.18 mg per year, which is significantly lower compared to the rest of the water area. For the eastern and central regions of the estuary, as well as the Kinburn Strait, where the rate of OP removal is high enough, the specific AC value was from 2.17 to 2.31 mg per year.

Consideration of the spatial and temporal changes in the content of OPs and phenols (sum) as priority indicators of water quality in the waters of the Dnieper estuary and the port of Odessa for the period 1996–2006 was the first step towards assessing the ability of these ecosystems to self-cleaning and rationing of pollutant discharges. The data array analyzed in the paper [15] is presented by the results of observations of the content of pollutants: in the waters of the Dnieper estuary—2.2 thousand definitions, in the waters of the port of Odessa—3 thousand definitions. In 1996–2006, in the waters of the Dnieper liman, the maximum content of OP and phenols exceeded MAC 28 and 22 times respectively (MAC_{OP} 0.005 mg/dm^3, $MAC_{phenols}$ = 1 μg/dm^3). During the period of research in this area, 80–83% of the total number of OP definitions reached or exceeded MAC. Excess of MAC for phenols was noted in 17% of the total number of definitions. In the sea waters of Odessa port during this period, the maximum OP and phenol concentrations reached 18 and 28 MACs respectively. The OP content in the surface waters of the port water area was constantly above the maximum allowable value, in bottom waters OP concentrations below MAC were recorded only in 9% of the total number of definitions. Excess of MAC in phenols was noted for 86% of the total number of definitions [16].

Schemes of location of hydrochemical monitoring stations in the water area of the Dnieper estuary and the port of Odessa in 1996–2006 are presented in Fig. 10.2.

Assessment of AC values of the Dnieper estuary and Odessa port ecosystems was carried out according to the methodology considered in [16, 17] based on the assumption of the spatial homogeneity of the fields of pollutants distribution within their boundaries. An important stage in the calculation of AC is the estimation of the time of removal of pollutants from the ecosystem. This is a rather complicated process. Its implementation requires either data from field studies and experiments in mesocosmos or from environmental modelling. In the paper [7] a new statistical method for solving the problem was proposed, applicable to the areas of state hydrochemical monitoring.

The balance sheet method was used to calculate AC which traditionally refers to the most polluted waters of Odessa Commercial Seaport, located off the south-western coast of Odessa Bay in the northwestern part of the Black Sea and is the largest seaport in Ukraine. Under the conditions of complex morphometric features, the quality of sea waters in the port water area is formed under the influence of the main longshore drift containing pollutants supplied by the flow of the Dnieper and South Bug Rivers, which are pressed by intensive navigation and functioning of the city and port infrastructure. One of the priority pollutants, as for the overwhelming majority of port waters, is oil products (OP), the content of which during the period of studies of 1996–2006 in the water area of Odessa port on average more than three times exceeded the maximum allowable concentration (MAC = 0.05 mg/dm^3), in extreme cases reaching 15–18 MAC. According to the monitoring data of the mentioned period, the self-cleaning capacity of the sea area was assessed. "So, the average specific AC of the port ecosystem (here and hereinafter per unit of

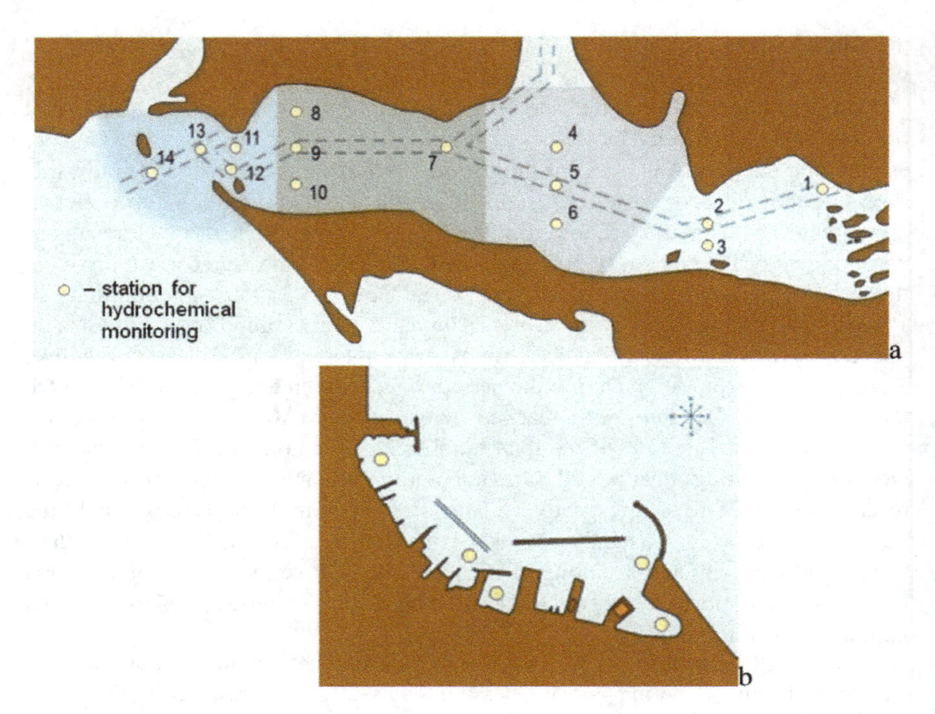

Fig. 10.2 The location of stations for hydrochemical monitoring in the Dnieper estuary (**a**) and the port of Odessa (**b**) in 1996–2006

volume equal to 1 dm^3) in relation to OP was 2.6 mg/year" [16]. Further, taking into account the significant amplitude of seasonal variations in sea water temperature and the associated activity of oil-oxidizing microflora, a detailed assessment of the AC of the marine ecosystem of the Odessa port water area in respect to oil products was made by studying its seasonal variations [18]. According to the data [18] "Seasonal changes in the level of marine water pollution in general during the period of observation was characterized by the maximum content of OP in the summer-autumn period, when the average monthly content was 4 times higher than MAC". Increased oil pollution of surface waters in April (more than 5 MACs) was formed by the longshore drift of flood waters of the Dnieper and the Southern Bug. The minimum oil content was in January–February, the average value in these months was about 2 MACs [7].

A database of the daily monitoring system is a prerequisite for carrying out work at this stage. The influence of changes in the average monthly temperature of sea waters on the rate of OP removal is shown in Fig. 10.3.

It was established [18] that "specific AC value of the water area of the port of Odessa for the warm season in the period 1996–2006 averaged $7.8 \cdot 10^{-3}$ mg per day, for the cold season in the period 1997–2006—$5.9 \cdot 10^{-3}$ mg per day. Thus, during the warm season the marine ecosystem is able to remove 0.07 t of OP per day more than during the cold period of the year without damage. Consequently,

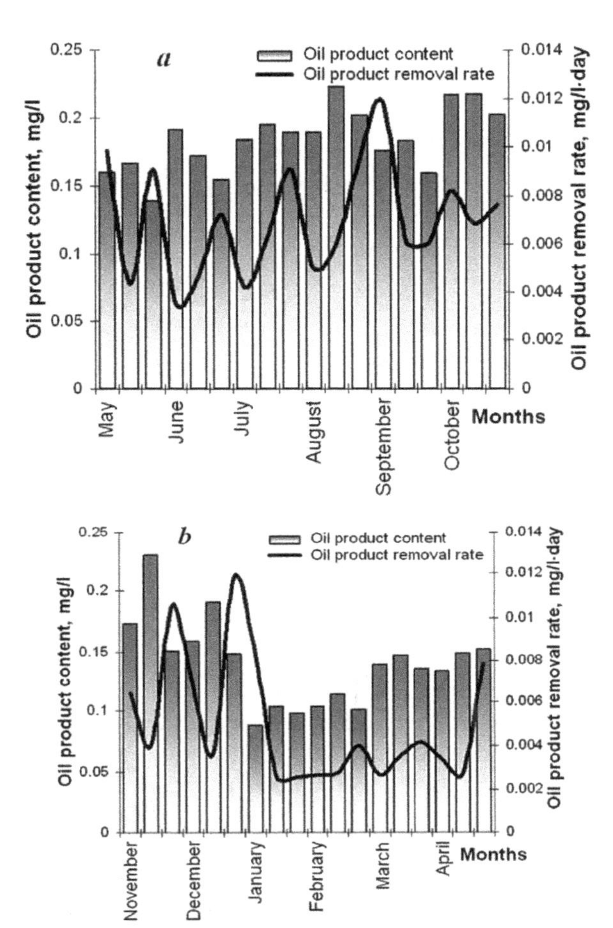

Fig. 10.3 Variation of the oil product content in the waters of the port of Odessa and the oil product removal rate during warm (**a**) and cold (**b**) seasons [18]

the decrease in the self-cleaning capacity of the marine ecosystem during the cold season indicates its greater vulnerability during this period of the year and should be accompanied by stricter pollution control".

The change in specific OP elimination rate in different regions of the Dnieper estuary was determined primarily by the intensity of hydrodynamic processes. Thus, in the water area of the estuary, on average, for the period 1996–2006, the highest rate of OP removal was typical of the central part (Table 10.1), where the discharge of river waters of the Southern Bug E.E. is imposed on the Dnieper River runoff [15].

In the eastern part of the estuary, the dynamics of which is determined by the Dnieper water inflow, the average rate of OP removal was less. The western part of the estuary, on the one hand, is locked by the narrowness of the Kinburn Strait and acts as a lagoon, and, on the other hand, is exposed to high anthropogenic load associated with intensive shipping (especially in its northern part, where the port

Table 10.1 Characterization of the rate of OP and phenols removal (sum) from the ecosystem of certain areas of the Dnieper estuary and the water area of the port of Odessa for the period of studies 1996–2006

Research area	OP removal rate mg/(dm^3-year)	Phenol removal rate (sum) µg/(dm^3-year)
Eastern region	2.31 ± 1.15	–
Central area	3.32 ± 1.39	–
Western region	1.80 ± 0.58	–
Kinburn Strait	3.10 ± 1.55	–
Dnieper estuary	2.38 ± 1.01	–
Port Odessa water area	2.42 ± 1.52	31 ± 9

city of Ochakov is located), was characterized by a minimum (average for the period of research) rate of OP removal on the background of high contamination of estuaries.

Estimated time of OP stay in the waters of the Dnieper estuary showed annual 8-fold flushing by river waters of water areas of the eastern and central regions. The maximum time of OP stay (71 days) was typical for waters of the western estuary. The renewal of waters in the Kinbur Strait and the prefluve part of the north-western Black Sea shelf took place on a monthly basis.

As can be seen from Table 10.1, on the whole, "during the period of studies of 1996–2006 the specific rate of OP removal was comparable for the water area of Odessa port. In terms of annual average values, the specific rate of phenols elimination (sum) varied from 18 to 45 µg/(dm^3-year), OP—from 1.11 to 5.83 mg/(dm^3-year)" [15].

Knowing the time of removal of pollutants from the ecosystem allows us to calculate the specific value of AC to normalize the permissible release of pollutants in each of the studied areas.

The average specific value of AC for the entire Dnieper estuary ecosystem with respect to the OP can be estimated at 1.94 mg/(dm^3-year), for the adjacent part of the Black Sea at 2.74 mg/(dm^3-year). For the eastern and central regions of the estuary and the Kinburn Strait, where the rate of OP removal is high enough, the specific value of AC was from 2.17 to 2.31 mg/(dm^3-year). For the strait of the western estuary, which is locked by the narrowness of the strait, it is much lower (Table 10.2).

For AC in the marine ecosystem, not only the rate of Sound removal is crucial, but also the volume of the ecosystem as a whole. For example, the central part of the Dnieper estuary water area had the maximum self-purification capacity for OP during the research period (Table 10.2). The average value of the assimilation capacity of the western region with a commensurate ecosystem volume is half as low due to the minimum rate of OP removal. The ecosystem of the Kinburn Strait was characterized by the minimum self-cleaning capacity, because even at a high rate of OP removal this part of the estuary occupies the smallest volume. It is quite obvious that in a number of investigated areas, at a comparable high rate of OP

Table 10.2 Characteristics of the assimilation capacity of the ecosystem of selected areas of the Dnieper estuary and the water area of the port of Odessa with respect to the OP for the research period 1996–2006

Research area	Specific value of assimilation capacities, mg/(dm^3-year)	Assimilation capacity of the ecosystem, thousand tons per year
Eastern region	2.17	2.36 ± 1.13
Central area	2.31	3.36 ± 1.50
Western region	1.18	1.72 ± 0.60
Kinburn Strait	2.28	0.26 ± 0.13
Dnieper estuary	1.94	7.96 ± 2.93
Odessa water area	2.60	0.0935 ± 0.0614

extraction, the minimum self-cleaning ability of the ecosystem of the water area of the port of Odessa is explained by a much smaller volume of water area [15].

Taking into account the fact that the limits of variation in the quantitative evaluation of AC are quite large (Table 10.2), in order to prevent further degradation and restore the gene pool of the areas under consideration during normalization of pollutant discharges into the water area as a "threshold" value, it is advisable to focus on the minimum AC value. For the Dnieper estuary, it is 5 thousand tons per year for OP, and 32 tons per year for the Odessa port water area for OP. At the same time, when considering the calculated AC value in comparison with the balance estimate of OP for the Dnieper estuary water area, it is indicative that at present, their inflow only with the river waters of the Dnieper and Yuzhnyi Bug (through the Bug estuary) makes 7 thousand tons of OP per year. Thus, the calculated value of AC is much lower than the incoming OP balance item, which negatively affects the state of the reservoir with fishery value.

Moreover, it should be remembered that the creation of favorable conditions for the development of marine ecosystems could be judged only under conditions of uniform application of pollutants in the amount not exceeding the value of AC. In case of heterogeneous inflow of pollutants, the threat to the ecosystem is posed by sources, which create a load exceeding the specific value of AC. Therefore, the specific value of AC cannot be comparable with salvo accidental pollution of the marine ecosystem at a local point. However, for example, in 1997 alone, 0.987 ton OP was discharged into seawater in the port of Odessa as a result of an accident at the Monte Chiaro mills [19]. At the same time, the average AC of the entire port ecosystem even per day was 0.09 t OP. Such a significant excess of AC could certainly lead to irreversible degradation of the biotic component of the ecosystem.

10.4 Assessment of the Self-cleaning Capacity of the Ecosystem of the Water Area Adjacent to the Heracleian Peninsula (Crimea)

The Heracleian Peninsula is a triangular land protrusion into the Black Sea in the south-western part of the Crimean Peninsula, which is separated from it by the Sevastopol Bay, the Black River, the Balaklava Valley and the Balaklava Bay. Its coastal water area is under the indirect influence of all the Sevastopol bays (Sevastopol, Krugla, Streletskaya, Cossack, Kamyshova and Balaklava). Anthropogenic load on this water area includes: releases of domestic and industrial wastewater, river runoff, areas of recreational load along the coast. In the studied water area there is a main sewerage discharge of Sevastopol (more than half of all municipal discharge). The pipe is laid on the bottom so that the diffuser is at a distance of 3.3 km from the coast. The powerful inflow of nutrients and dissolved organic matter leads not only to a sharp increase in phytoplankton production, but also to changes in the hydrochemical regime of this water area. According to satellite data, in the coastal area between the m. Chersonesos and m. Fiolent surface anomalies of temperature and salinity were repeatedly observed, as well as spots of unknown origin, which could be caused not only by hydrological (upwelling), but also anthropogenic (wastewater discharge) factors. In this context, it was of considerable interest to identify coastal areas with anomalous values of hydrochemical characteristics using the analysis of their variability as a "background". The work [20] analyzed an array of historical data of 2025 hydrological stations and 337 hydrochemical stations of the water area near the Heracleian peninsula for the period 1960–2010 from the oceanographic data bank of the FSBUN MHI RAS.

The investigated water area is also subject to recreational load with distinct seasonality, which may manifest itself in the seasonal variability of oil products content at the site. As part of the research on the ecological state of the water area of the Heraclean peninsula in the spring and autumn periods of 2016, the content of oil products in the water area of the micro polygon was assessed and the assimilation capacity of its ecosystem was calculated using the synoptic method [21]. The scheme of stations of the micro polygon sampling for oil products is presented in Fig. 10.4.

According to the work [22] "in spring concentration low excess of MAC is insignificant about 1.2, 1.5 MAC, in autumn excess of MAC is more significant than 6 MAC". According to the authors "This situation is quite understandable—in autumn after the holiday season with increased recreational load accompanied by increased transport load on the water area and increased volume of sewage runoff, the oil content in the water area increases. In addition, during the holiday season increases the number of low-tonnage vehicles, the condition of which is almost uncontrolled, they are an additional source of oil hydrocarbons (OH) due to the leakage of fuels and lubricants.

The "synoptic" method was used to calculate the assimilation capacity of the ecosystem of the investigated water area. According to salinity and temperature measurements, one water mass was present at the micro polygon, its volume was at the surface layer depth of 0.5 and the micro polygon area of 1.2 km^2 = 0.6 km^3.

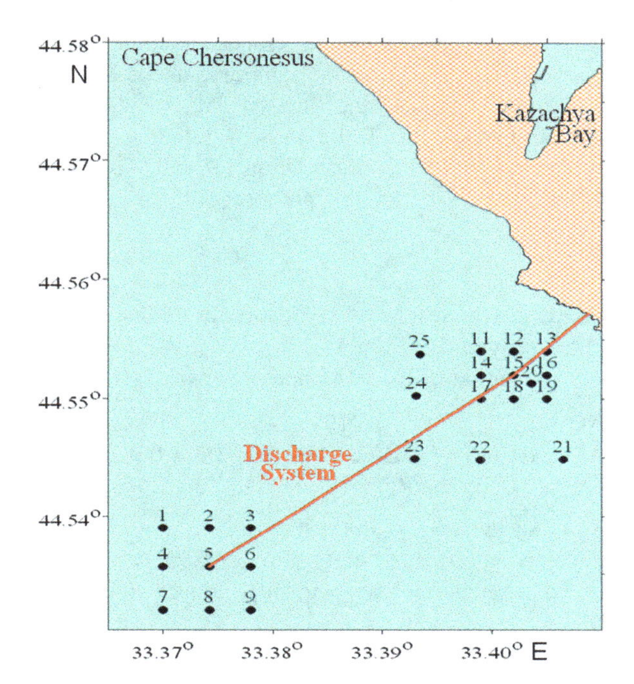

Fig. 10.4 Scheme of sampling stations for oil products (OP)

The difference between the maximum (C_{max}) and minimum (C_{min}) concentrations calculated for a specific pollutants and each water mass was used as an indicator of inhomogeneous distribution of pollutants. According to the synoptic AC calculation method, the heterogeneity is caused by self-cleaning processes, and the starting point for these processes is the date of the last storm. In the paper [21] this date was conditional, denoted as N and was considered in several variants: 2, 5,10, in the present work the variant where N = 2 days was calculated.

The assimilation capacity is calculated using a formula:

$$\mathrm{AC} = [(\mathrm{C_{max}} - \mathrm{C_{min}})/\mathrm{N}] \cdot C_{\mathrm{ma}}/\mathrm{C_{max}} \tag{10.1}$$

The AC dimension in this case corresponds to the pollutant concentration dimension in water. If the concentration is expressed in µg/L, the AC dimension is µg/L•per day. In order to calculate AC of the investigated water area, information on water volume of the investigated area is necessary and is calculated for a calendar year (365 days).

In addition to the assimilation vessel, the pollutant load can be calculated using a formula:

$$\mathrm{H} = \left(\mathrm{C_{max}} - \mathrm{C_{average}}\right)/\mathrm{N}\ (\text{dimension: mg/l} \cdot \text{per day}) \tag{10.2}$$

At H $\geq$ AC, the conclusion is made about the ecological disadvantage of the water mass, in H $\leq$ AC, on the contrary, about the ecological well-being. It is also possible to calculate the time (t, day) required by the ecosystem to assimilate, without harming itself, the mass of pollutants to the level of MAC.

Accordingly, this calculation is only possible if the $C_{maximum} \geq$ MAC. A formula is used for this purpose:

$$t = (C_{max} - C_{MAC})/AC \tag{10.3}$$

Below are the results of calculation of all listed parameters for spring 2016 for the micro polygon.

Calculation of the assimilation capacity (A E) on the micro polygon in spring

AC = [(C_{max} − C_{min})/ N] · C_{MAC}/ S_{max} − variant, where N is equal to 2 days.
AC = [(0.07 − 0.012)/2N] · 0.05/0.07 = 0.0205 mg/l per day.
AC = 0.0205 mg/l per day or 7.482 mg/l per year

or 4,488 tons per year for the total volume of surface water layer (0.5 m) of the micro polygon under study (0.6 × 10^6 m^3).

To calculate the environmental impact on the water area

$$\text{Load H} = \left(0.07C_{max} - C_{average}0.0275\right)/2N \text{ (dimension: mg/l} \cdot \text{daily)}$$

H = 0.00096 mg/L per day[22]. On the micro-test site H $\leq$ AC 0.00096 $\leq$ 0.0205 that testifies to ecological well-being on the test site in spring. Thus, in spring the ecological situation at the microware is favorable in contrast to the autumn period, the calculation for which is given below.

The calculation of the assimilation capacity (AC) on the micro polygon in autumn was carried out according to formula (10.1)

$$AC = [(0.340 - 0.001)/2N] \cdot 0.05/0.340 = 0.02491 \text{ mg/l per day.}$$
$$AC = 0.02491 \text{ mg/l per day or } 9.094 \text{ mg/l per year}$$

or 5,456 tons per year for the whole volume of surface water layer (0.5 m) of the investigated micro polygon (0.6 × 10^6 m^3) [22].

To calculate the environmental impact on the water area by formula (10.2)

$$\text{Load H} = \left(0.340\ C_{max} - C_{average}0.0806\right)/2N \text{ (dimension: mg/L} \cdot \text{daily)}$$
$$H = 0.1297 \text{ mg/l per day}$$

At the micro polygon in the fall of 2016. H $\geq$ AC 0.1297 $\geq$ 0.02491, which indicates an environmental poor condition at the site in the autumn, which allows the formula

$t = (C_{max} - C_{MAC})/AC$ (3) to calculate the time (t, day) required by the ecosystem to assimilate, without damage to itself, the mass of pollutants in our case to the level of MAC. For the autumn of 2016 this time was equal:

$$t = (C\ 0.340 - C\ 0.05_{MAC})/AC\ 0.02491 = 11.6\ \text{days}$$

Thus, calculations have shown that the higher the concentration of oil products exceeds the MAC, the longer it will take the ecosystem to assimilate the pollutant.

According to the authors [22] “activation of mobile fishing and recreational services during the holiday season contributes to the increase in the number of low-tonnage vehicles, the condition of which is practically uncontrolled, they are an additional source of oil hydrocarbons (OH) due to the leakage of fuels and lubricants”.

10.5 Self-cleaning Capacity of the Sevastopol Bay Water Area Ecosystems Depending on the Level of Anthropogenic Load

S1evastopol Bay is an estuarine seashore of the Chernaya River—a closed estuary-type water area with limited water exchange, which is under constant technogenic impact (shipping, docking, hydraulic works) and anthropogenic influence (domestic urban and rainwater runoff). On the one hand, the bay is a town-forming element of the geosphere, and on the other hand, it serves as a base for the Russian Navy with all the industrial, production and economic infrastructure typical of such complexes, as well as a zone of active navigation and hydrotechnical works, which leads to disturbance of existing equilibria. This situation is significantly aggravated by the fact that water exchange between the bay and the open sea is complicated by the construction of a protective breakwater in 1976–1977. A significant influence on the formation of the hydrochemical structure of the bay’s waters is produced by the river (in the eastern part of the bay it receives the waters of the Black River) and terrigenous runoffs, as well as domestic urban runoffs, with which the bay receives additional biogenic elements, which are the material basis of the biotic cycle and a fundamental factor in the process of eutrophication of the water body.

Urban industrial enterprises, economic and recreational facilities located on the adjacent territory impose a high uneven load on the sea area. The scheme of location of the main sources of pollution in the bay is presented in Fig. 10.5 [23].

The regionalization of the Sevastopol bay in terms of morphometric characteristics was carried out in the work [24], which coincided well with the pollution zones allocated in the work [23].

Depending on the localization of pollution sources, morphometry and hydrometeorological conditions, both relatively “clean” zones and zones of

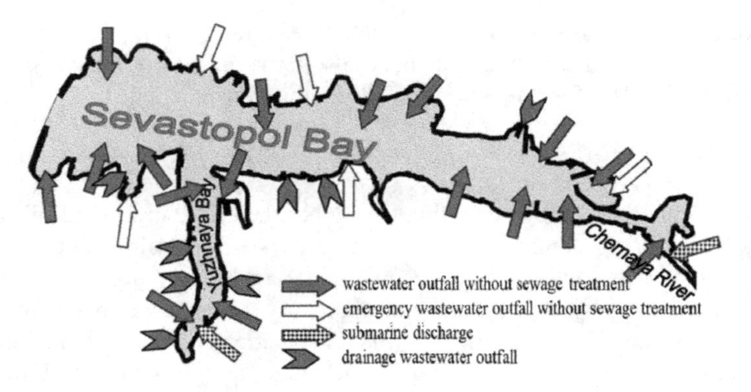

Fig. 10.5 Sources of water pollution of the Sevastopol Bay

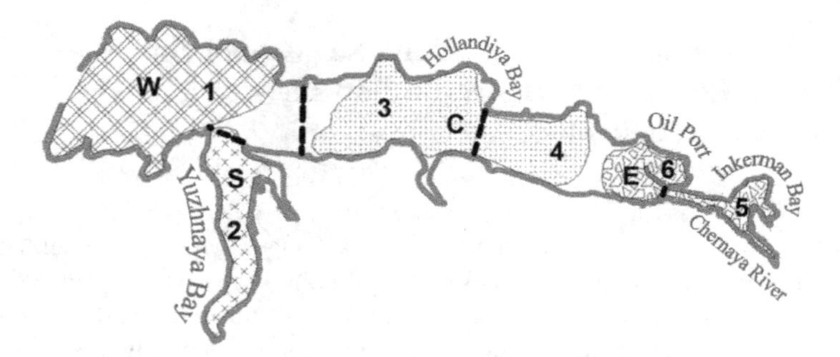

Fig. 10.6 Zoning of the Sevastopol Bay by the water pollution levels [25]

sustainable high pollution levels (e.g., the Southern Bay) are formed in the Sevastopol Bay. In [25] the water area of the Sevastopol Bay was divided into four areas. Geographically, the zone of low pollution is the western zone "W", the moderate zone is the eastern zone "E". The zone of heavy pollution occupies the central part of the bay—"C". The southern bay, "S", is very heavily polluted (Fig. 10.6).

In Fig. 10.6, numerals indicate areas by Stokozov[24], letters by Ivanov et al. [25]. Calculated in [24] morphometric characteristics (volumes of layers, surface areas, cross-section areas, lengths of coastlines and boundary lines) for the Sevastopol Bay and its isolated parts allow to calculate the reserves of chemicals in the water and bottom sediments of the bay, and with knowledge of flows of intake and removal of substances to obtain estimates of the assimilation capacity of ecosystems of these water areas.

In this section of the book, based on the data of long-term monitoring using the calculated values of assimilation capacity (in the implementation of two methods (balance and synoptic), as well as the calculation of the trophicity index, and taking into account the peculiarities of the hydrodynamic regime of the bay and the impact of runoff of the Chernaya River, a comparative analysis of the ecological state and self-cleaning capacity of all parts of the Sevastopol Bay selected by the level of anthropogenic load.

For Sevastopol Bay ecosystems, the assimilation capacity (AC) was calculated in relation to inorganic forms of nitrogen as a priority pollutants (PP) in municipal and stormwater runoff, and is the material basis of the biotic cycle and a fundamental factor in the eutrophication process.

The paper [26] considers the implementation of two methods for quantitative determination of self-cleaning ability of marine ecosystems (balance and synoptic). Based on the materials of long-term monitoring studies, a comparative analysis of the assimilation capacity of the most ecologically unfavorable part of the Sevastopol Bay (Yuzhnaya Bay) with a cleaner part of its water area bordering the open sea in relation to inorganic forms of nitrogen as priority pollutants in municipal and stormwater runoffs was carried out. Environmental well-being of the studied areas of the Sevastopol Bay water area, both in respect of inorganic forms of nitrogen and oil products, was assessed.

The work was based on the results of monitoring observations of the Marine Hydrophysical Institute of the Russian Academy of Sciences (MHI RAS) carried out in the waters of the Sevastopol Bay at 36 stations over the period 1998–2012 (scheme of the stations in Fig. 10.7). The information on the content of inorganic forms of nitrogen (nitrates, nitrites, ammonium) in sea water from the MHI oceanographic data bank was used in the work.

Comparative analysis of self-cleaning ability with respect to forms of mineral nitrogen was carried out for the most environmentally disadvantaged part of the Sevastopol Bay (Yuzhnaya Bay—District S) and a cleaner part of its water area, bordering the open sea (District W). The choice of the most polluted and cleanest part of the water area of the Sevastopol Bay is based on the results of zoning of the Bay of Sevastopol on the level of anthropogenic load [25] and on the features of its

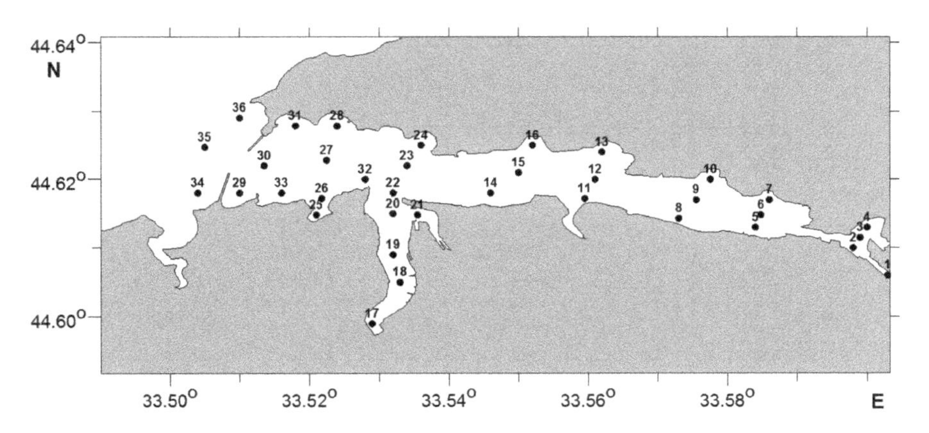

Fig. 10.7 Scheme of hydrochemical monitoring stations of the Sevastopol Bay

morphological structure [24]. In the bay there are areas of weak, moderate, strong and very strong pollution (Fig. 10.6).

For the Southern Bay (District S, Fig. 10.5) and part of the Sevastopol Bay adjacent to the open sea (District W, Fig. 10.5), the self-purification capacity in relation to inorganic forms of nitrogen (nitrites, nitrates and ammonium nitrogen) was assessed as a priority pollutant in municipal and storm sewerage runoffs, as well as active in the production and destruction processes of the marine ecosystem. To calculate the AC of the ecosystem of Area S, an array of data on the three forms of nitrogen (Table 10.3) was used, which made up 714 definitions, and for Area W —1117 definitions obtained during the time interval 1998–2012.

To diagnose the ecological well-being of the water area taking into account AC, the *PI* index was used, which characterizes the degree of load deviation for the *ith* component per year from *ACi* and calculated by the formula [29]:

$$PI = \frac{(C_i - MAC_i) \cdot V_i}{M_i} = 1$$

where C_i is the average content of ith pollutants, V_i is the water volume in the calculated area, M_i is the calculated permissible AC mass of matter for the water area per year.

According to [29] the state of the system was considered as safe if $P_i \leq -1$. In this case, $C^i \leq \mathrm{MAC_i}$, the level of pollution does not exceed the permissible. At a relative prosperity of $-1 \leq P_i \leq 0$, the initial level of pollution is leveled by self-cleaning processes to an acceptable level. The state of the ecosystem with $P_i > 0$ is environmentally unfavourable. Loads on the ecosystem that exceed its self-cleaning capacity disturb the normal functioning of the system.

The southern area of the Sevastopol Bay, which includes Yuzhnaya Bay and Kilen Bay, is characterized by difficult water exchange with the main water area. In view of limited water exchange with the main water area and as a traditional location of numerous ship berths, on the volume of industrial, domestic and stormwater runoff the Southern Bay takes first place among other Sevastopol bays. Its cubic part is characterized by distributed areas, while the intensity of water

Table 10.3 Characteristics of the database on inorganic forms of nitrogen for selected areas of the Sevastopol Bay

Characteristics, units of measure	NO_{2-}		NO_{3-}		NH_{4+}	
	District S	W District	District S	W District	District S	W District
Number of definitions	240	373	225	351	249	393
Threshold of concentration, μg/L	1.43		221.43		20.71	
Average content, μg/L	0.23	0.12	12.59	2.45	0.95	0.57
Maximum concentration, μg/L	1.48	0.42	142.79	13.31	8.17	8.18

distribution is not constant throughout the year. This is explained by the fact that the south-western coast of the Sevastopol Bay is adjacent to a developed network of fracture zones in the basin of the Black Sea, as well as the Sardinaki beams, which is able to drain groundwater flow from large areas with significant groundwater reserves. Their significant share in the submarine discharge process goes to the Southern Bay, which is recorded by lower salinity and higher silica acid content [30].

Surface waters of the Southern Bay (especially in the cut-off part) are characterized by maximum concentrations of nitrogen compounds (Table 10.3). According to the information given in the work [31], untreated storm sewage and untreated sewage are discharged into the South Bay water area, and there are ship stops. At the same time, it should be noted that the average content of all forms of inorganic nitrogen during the observation period did not exceed the respective MACs, the values of which were used as a threshold value.

Comparative AC values of the South Bay (region S) and the purer region W for inorganic nitrogen forms are presented in Fig. 10.8. When calculating AC of the bay water areas under study, morphometric parameters data were used [24].

As follows from [26] and Fig. 10.8, "the water area of the Southern Bay (District S) is characterized by higher values of maximum concentrations compared to the purer water area (District W). As for nitrates, the excess is more than one order of

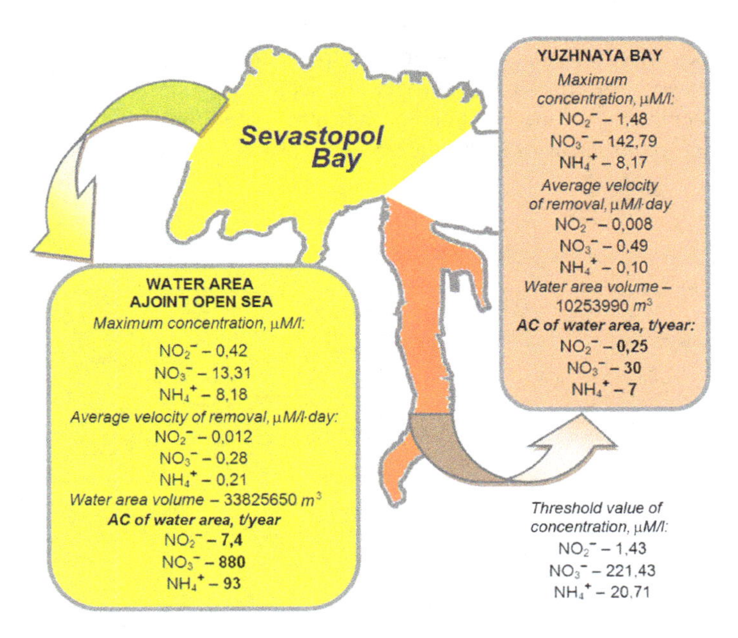

Fig. 10.8 The balance method for calculating the AC of ecosystems of individual regions of the Sevastopol Bay with different levels of anthropogenic load in relation to inorganic forms of nitrogen

magnitude, for nitrites—more than three times, and only for ammonium its content in the studied water areas is within the same limits".

"Maximum specific removal rates during the study period for the Southern Bay reached 0.03 μg/L per day for nitrites, 2.10 μg/L per day for nitrates, and 0.32 μg/L per day for ammonium nitrogen, exceeding the average removal rates of 3.5–4 times. Accordingly, the maximum specific removal rates for Area W were 0.066 μM/L per day for nitrites, 1.28 μM/L per day for nitrates, and 0.94 μg/L per day for ammonium. For cleaner water area the excesses of average removal rates were from 4.5 to 5.5 times" [26]. Apparently, this is due to the predominance of biological processes in cleaner water areas (recirculation of inorganic forms of nitrogen) over anthropogenic processes, but an additional research is required to ascertain the reasons.

The South Bay ecosystem's self-cleaning capacity has been estimated to be no more than 0.25 tons of inorganic nitrogen entering the study sea area per year for nitrites, 30 tons for nitrates and 7 tons for ammonia nitrogen. It should be noted that the shown quantitative restriction of inorganic nitrogen discharges into the water area is allowed only if the pollution is uniform and planned. In cases of emergency volley discharges in assessing the ability to self-clean this ecosystem should be oriented to the specific value of AC, which is a component for nitrites 0.0048, for nitrates 0.58 and for ammonium nitrogen 0.13 μm/L per day. How many times this level is currently exceeded in the South Bay, or is below it, is difficult to estimate as there is no accurate data on the total amount of inorganic nitrogen entering the specified water area with storm and municipal runoff. For area W, AC is higher and the limit load may be 7.4 tons per year for nitrites, 880 tons per year for nitrates and 93 tons per year for ammonium nitrogen, respectively.

The calculation of the well-being indicator based on 1998–2012 averages showed that for all forms of inorganic nitrogen the ecosystems of the selected areas fall into the zone of sustainable well-being (Fig. 10.9). Load on the considered

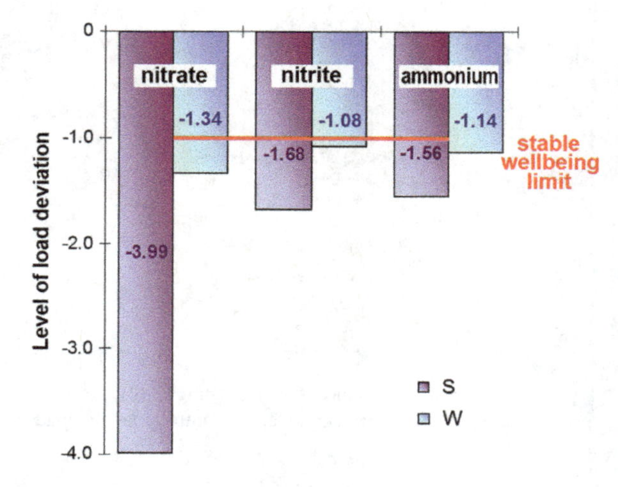

Fig. 10.9 The indicator of the ecological well-being of certain regions of the Sevastopol Bay with different levels of anthropogenic load in relation to inorganic forms of nitrogen

areas does not exceed its self-cleaning capacity and, accordingly, does not disturb the normal functioning of the system.

Tail end eastern part of the Sevastopol bay, which includes the mouth of the Black river, navigable canal, Inkerman bucket and Neftegavan, shallow, according to the work [23] its average depth is 4.7 m, and the maximum—10.8 m. The water area is polluted due to sewage discharge (without treatment) and emergency releases in the area of Neftegavan, thermal power plant (TPP) and at the mouth of the Chernaya River, where there are also sources of submarine discharge.

The studied part of the Sevastopol Bay, as the most ecologically vulnerable part of the bay, is significantly affected by winter-spring floods on the Chernaya River.

According to the data [32] distributed flood waters (February 2015) affected the level of nutrient inputs into the bay water area, increasing the content of inorganic forms of nitrogen and silicates, which should certainly be taken into account when calculating the assimilation capacity of the ecosystem of the investigated water area with respect to inorganic nitrogen and estimation of the trophicity index.

An array of analyzed field data on the eastern part of the Sevastopol Bay is presented by the results of 750 determinations of inorganic nitrogen content in sea waters, performed by photometric method in the stationary chemical laboratory of the MHI RAS. The scheme of water sampling stations location in the eastern part of the Sevastopol Bay and in the Black River is presented in Fig. 10.10.

In spite of the significant nitrogen inflow with runoff of the Chernaya River [33], in the eastern part of the Sevastopol Bay the average for the period of 1998–2012 was much lower than the corresponding MAC, which meets the conditions for using the balance method of AC assessment of the marine ecosystem.

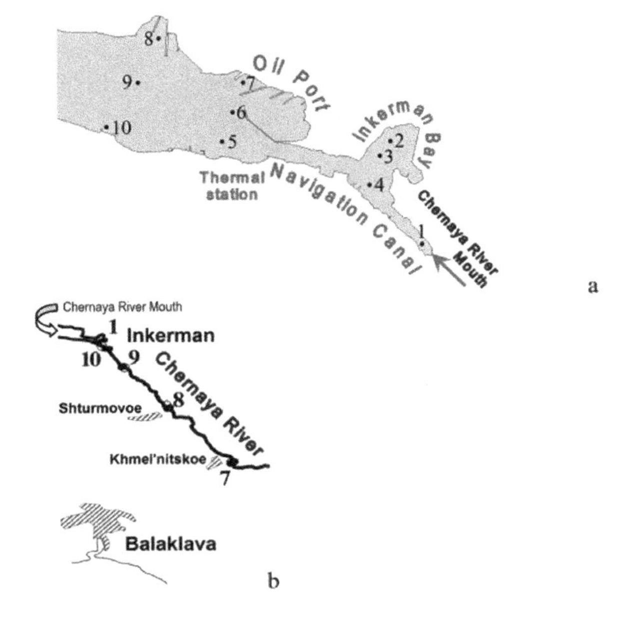

Fig. 10.10 Schemes of sample stations in the Sevastopol Bay apex (**a**) and in the Chernaya River (**b**)

To determine the rate of removal of inorganic nitrogen from the study area was chosen time series containing 55 values of nitrates, 56 nitrites and 49 ammonium concentrations. Estimated removal rates varied in a wide range, the maximum daily values reached 0.781; 0.015 and 0.194 μM/L for nitrates, nitrites and ammonium, respectively [35].

Estimated time of stay of mineral complexes of nitrogen in waters of investigated water area has shown, that nitrates and ammonium are deduced from an ecosystem on the average for month (28 days for nitrates and 34 days for ammonium). Removal of nitrites occurs much slower, on the average more than 50 days are required. Presented in [33] the calculation of nutrients removal by the closing shaft, located at a distance of 11 km from the mouth of the river Chernaya (Khmelnitskoe), showed that the Sevastopol bay annually receives about 20 tons of mineral nitrogen with the river flow on average. Moreover, given the fact of significant water transformation in the lower reaches of the river, the actual nitrogen supply may exceed 50 tonnes. It should be added that according to [33], the contribution of anthropogenic sources in the supply of biogenic substances to the Sevastopol Bay is up to 52% of the total supply of mineral nitrogen. Thus, a comparative analysis of the AC value of the ecosystem in the eastern part of the Sevastopol Bay with inorganic nitrogen inputs only from the waters of the Chernaya River (according to the data [33]) shows a significant excess of pollutant flow over the capacity of the receiving water area for self-cleaning (Table 10.4).

Calculation of the AC value of the water area ecosystem of the eastern part of the Sevastopol Bay was based on the assumption of spatial homogeneity of the fields of biogenic elements distribution within its boundaries.

At the same time, despite the significant imbalance between the introduction of biogenic complexes with river waters and the calculated capacity of the water area for self-cleaning, it is important to take into account their uneven inland flow into the bay water area, i.e. seasonality of their involvement in biological processes. Thus, according to [34], in the waters of the Chernaya River entering the bay the dominant form of inorganic nitrogen is nitrate, the minimum concentrations of which occur during the vegetation period (April–June). The maximum content of the sum of mineral nitrogen compounds ΣN (NO_3^- + NO_2^- + NH_4^+) is typical for autumn-winter floods and can also be traced in the intermittent period. And if

Table 10.4 Comparative characterization of the ecosystem capacity of the eastern part of the Sevastopol Bay for self-cleaning with respect to forms of inorganic nitrogen during the period of studies in 1998–2012 and their transfer to the Sevastopol Bay with the river flow [35]

District	AC ecosystems in terms of inorganic nitrogen forms, tons per year		
	NO_2^-	NO_3^-	NH_4^+
Sevastopol Bay	0.18	17.36	3.75
Khmelnitskoye, closing wing of the Chernaya River	0.17	9.5	9.4
Lower reaches of the Chernaya River	0.23	42.9	7.5

during the flood period a significant inflow of nitrogen can to some extent be compensated by the intensification of the dynamic transport outside the water area under consideration, then during the summer low-water period the high content of biogenic complexes can be accompanied by blooming of sea waters with the subsequent development of hypoxia and overseas phenomena.

The central part of the Sevastopol Bay as a whole is the deepest, has an average depth of 13.1 m with maximum values in the fairway up to 19.5 m, is located at a significant distance from the mouth of the Black Sea river, wastewater and emergency discharges in the Southern Bay, in the area of Neftegavan, thermal power plant (TPP) and in the eastern part of the Tail end Bay, where there are also sources of submarine unloading.

The ecological state of the central part of the water area of the Sevastopol Bay is under the indirect cumulative influence of all coastal sources of pollution, runoff of the Chernaya River, especially in the periods of flooding, and is determined by the dynamics of water and thermohaline regime, which depend on climatic characteristics of the coastal zone of the south-western part of Crimea, wind regime, changes in water exchange conditions with the sea and inside the bay [31]. Along the coast of this part of the Sevastopol Bay, there is a much smaller amount of stormwater and sewage discharges without treatment than in other parts of the Bay (e.g. the Southern Bay) (see Fig. 10.5).

According to [25] waters of the central part of the Sevastopol bay are characterized as dirty, but by the total inorganic nitrogen content on average for the period 1998–2012 the waters of the region were one and a half times cleaner than the adjacent water areas and 5.4 times cleaner than the extremely polluted water areas of the Southern bay (Fig. 10.5).

The average values of all forms of inorganic nitrogen during the observation period did not exceed the respective MACs, the values of which were used as a threshold value, which meets the conditions of the application of the AC balance method of the marine ecosystem assessment as for the eastern part of the bay.

The presence of two counterflows within the water area—one from east to west from the Chernaya River and the second from the open part of the sea—contributes to the formation of a buffer zone in the central part of the Sevastopol Bay, which as if closes the multi-directional flows, including "enriched" by pollution. Substantially lower levels of inorganic forms of nitrogen for the Central part of the bay in all seasons are shown in [8].

The characteristics of the rate of mineral nitrogen removal and the maximum allowable volumes of its inflow into the water area of the central part of the bay are considered in work [36]. Specific (in terms of 1 litre) AC values obtained by calculation for each form of inorganic nitrogen differ from the corresponding MACs (Table 10.5), allowing a more accurate assessment of the limit of self-cleaning capacity of a particular ecosystem. The calculated removal rate varied widely, with maximum daily values reaching 0.12 μM/L for nitrates and nitrites and 0.20 μM/L for ammonium.

Table 10.5 Characteristics of the ecosystem capacity of the central part of the Sevastopol Bay for self-cleaning with respect to forms of inorganic nitrogen in the period of studies 1998–2012 [36]

Feature	Magnitude that characterizes the ability of an ecosystem to self-purify forms of inorganic nitrogen		
	NO_2^-	NO_3^-	NH_4^+
Average content, μM/dm^3	0.13	1.71	0.71
Removal rate, μM/(dm^3-year)	2.22	15.66	10.37
Removal time, h	555	1066	528
Specific (dm^3) AC value, μM/year	1.49	224.15	29.24

Despite the fact that the average for the period of observation rate of elimination of nitrates exceeds the corresponding indicator for nitrites and ammonium (Table 10.5), time, the estimated time of stay of mineral complexes of nitrogen in the waters of the study water area showed that nitrites and ammonium are removed from the ecosystem on average for 22–23 days. Removal of nitrates occurs twice slower, for full self-cleaning on the average more than 44 days are required.

According to the morphometric parameters of some parts of the Sevastopol Bay presented in the paper [24] and the assessment of the ability of the marine ecosystem under study to self-clean for the period of monitoring observations in 1998–2012, the amount of inorganic nitrogen entering the waters of the central part of the Sevastopol Bay during the year should not exceed 0.48 tons for nitrites, 71.56 tons for nitrates and 9.33 tons for ammonium nitrogen. It is necessary to pay attention that the shown quantitative restriction of inorganic nitrogen discharges is admissible only at their uniform (planned) input into the water area. In the conditions of emergency volley discharges or seasonal increase in the intake of biogenic complexes in assessing the ability to self-cleaning of the ecosystem under consideration should be guided by the specific value of AC, which is for nitrites 0.004, for nitrates 0.61 and for ammonium nitrogen 0.08 mM/L per day.

Calculation of the indicator of ecological well-being of the ecosystem of the central part of the Sevastopol Bay allowed to establish that for all forms of inorganic nitrogen, the ecosystem of the selected area falls into the zone of sustainable well-being (Fig. 10.11). Stress on the water area does not exceed its self-cleaning capacity and, accordingly, does not disturb its normal functioning.

Thus, analysis of inorganic nitrogen content in the waters of the Sevastopol Bay showed that the waters of its central part, located at a sufficient distance from typical sources of biogenic pollution, are characterized by the lowest total content of mineral nitrogen, mainly due to the nitrate component. Despite the largest for inorganic nitrogen components of the average rate of elimination of nitrates, the time required for their removal, twice the corresponding figure for nitrites and ammonium. At uniform (planned) receipt in the central water area of the Sevastopol Bay the amount of inorganic nitrogen for the year should not exceed 0.48 tons for nitrites, 71.56 tons for nitrates and 9.33 tons for ammonium nitrogen.

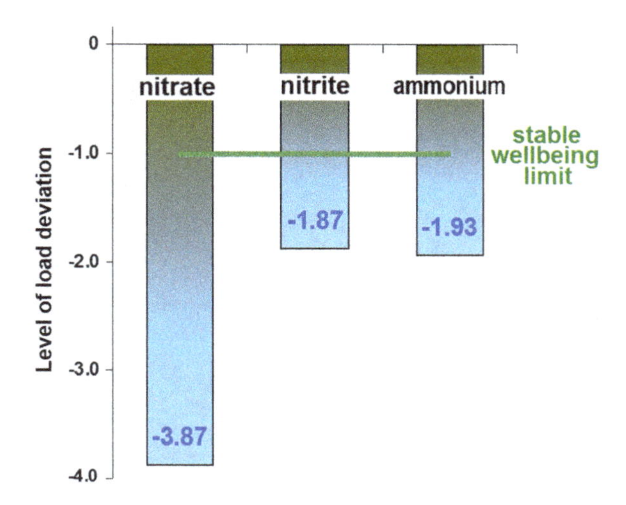

Fig. 10.11 The indicator of the ecological well-being of the ecosystem of the central part of the Sevastopol in relation to inorganic forms of nitrogen in 1998–2012

For the central part of the water area of the Sevastopol Bay, where there was a significant excess of the permissible level of oil pollution of waters according to the materials of complex oceanographic survey, the assimilation capacity, anthropogenic load and time of oil products assimilation by the ecosystem was estimated. AC values in relation to OP for the central part of the bay were calculated by the synoptic method [21] based on the survey data of October 2008. To implement this method it is enough to use integrated oceanographic survey data, which allows conditionally assessing AC of aquatic ecosystems not subject to state or departmental monitoring. However, AC assessment of an ecosystem not based on a number of long-term observations cannot characterize the basic self-cleaning ability of the water area, as this method is based on the assumption that inhomogeneous distribution of contaminants in water mass, which is homogeneous in terms of physical parameters, detected on the basis of survey results is a consequence of self-cleaning processes taking place in the water mass, the starting point for which is the last storm in the water area (therefore this method is called "synoptic"). For calculation of AC the results of analysis of OP content in water and bottom sediments of the Sevastopol bay at use of materials of complex survey in October, 2008, executed in laboratory of branch of the Moscow State University in Sevastopol on a technique [27] are used.

The results of the oceanographic survey of the bay in October 2008 included studies of oil pollution of waters at 16 stations on the surface horizon practically on the whole water area of the Sevastopol Bay. From the obtained data it follows that increased OP concentrations were observed in the central part of the bay (District C in Fig. 10.5). Maximum water pollution was 12 times higher than MAC (MAC = 0.05 mg/L) at concentrations ranging from 0.01 to 0.59 mg/L.

AC assessment of the ecosystem of the central part (region C) of the Sevastopol Bay in relation to the OP was performed by synoptic method. The significant disadvantage of this method is the dependence of AC assessment on the current hydrometeorological situation, which allows to conditionally assess the ecological state of the ecosystem only for the period of a specific survey.

The evaluation of AC using the synoptic method [21] is performed by the formula:

$$AC = [(C_{max} - C_{min})/N]\ C_{MAC}/C_{max}\ldots,$$

where AC—assimilation capacity (mg/L per day), C—corresponding to the concentration index of pollutants (mg/L), N—the period of time elapsed since the last storm (in our case 2 days). Load (H, mg/L per day) of pollutants per ecosystem can be calculated by the formula:

$$H = \left(C_{max} - C_{average}\right)/N$$

For H $\geq$ AC, the conclusion is made about the environmental degradation of the water mass, for H $\leq$ AC, on the contrary, about the environmental degradation.

It is also possible to calculate the time (t, day) required by the ecosystem to assimilate, without harming itself, the mass of pollutants to MAC levels. Accordingly, this calculation is only possible if $C_{max} \geq C_{MAC}$. A formula is used for this purpose:

$$t = (C_{max} - C_{MAC})/AC$$

The results showed the following values:

The AC of District C in relation to the NP made:

$$\begin{aligned} AC &= [(0.59 - 0.01)/2] \cdot 0.05/0.59 \\ &= 0.0243(\text{mg/l per day}) \text{ or } 8.869(\text{mg/l per year}). \end{aligned}$$

Using the work data [24] for the volume of surface water layer in the central part of the Sevastopol bay (V = 872,620 m^3) AC layer of 0–0.5 m was 7,739 tons per year.

The load (H) on the water area under study was equal:

$$H = (0.59 - 0.142)/2 = 0.224(\text{mg/l per day})$$

Since the load (H) exceeds AC, we can draw a conclusion about the environmental disadvantage of the studied part of the Sevastopol Bay water area with respect to the OP. It is also possible to calculate the time (t, day) required for the ecosystem to assimilate the mass of pollutants up to MAC level without affecting its functioning:

$$t = (0.59 - 0.05)/0.0243 = 22.2(\text{day})$$

Thus, the calculations as well as the work data [11] showed that the higher the OP concentration exceeds the MAC, the longer it will take for the ecosystem to assimilate the pollutant.

For District C with respect to OP, the well-being index (P) to [29] was −0.99. Despite the fact that the average OP content in Area C exceeds the corresponding MAC, a negative welfare indicator sign suggests that there is a reserve in the ecosystem's self-cleaning capacity. However, since the indicator value is in the range from 0 to −1, the ecosystem well-being is relative.

By the example of using two methods of AC calculation for marine ecosystems (balance and synoptic) it is shown that the synoptic method, implemented for the areas not covered by the monitoring studies, allows to conditionally assess the ecological state of the ecosystem only for the survey period. The balance method of AC assessment of marine ecosystems, taking into account their longer term functioning, allows, if necessary, to specify the AC value taking into account annual and/or seasonal changes in self-cleaning ability.

The steady ecological well-being of the water area in the central part of the Sevastopol Bay in relation to inorganic forms of nitrogen and relative to OP is shown.

10.6 Comparison of Self-cleaning Capacity of Ecosystems of Different Parts of the Sevastopol Bay Water Area by the Size of Their Assimilation Capacity and the Trophicity Index E-TRIX

Ecological condition of waters in the Sevastopol Bay is analyzed in comparison with the self-cleaning ability of ecosystems in the eastern, central, western parts of the Bay and the Southern Bay, using AC values in relation to inorganic nitrogen as a priority pollutants in municipal and stormwater runoffs, as well as the trophicity index E-TRIX with the use of biogeochemical parameters calculated using the one-dimensional version of the model of water quality Model for Estuarine and Coastal Circulation Assessment (MECCA) [37], taking into account the level of technogenic load and seasonality of biological processes (warm and cold period). This approach allows to adequately identify the areas most vulnerable to self-purification and formation of negative ecological situations flesh before the catastrophic.

To calculate AC ecosystems of different parts of the Sevastopol Bay, an array of data for 1998–2012 was used, which amounted to 4144 element-definition. The average values of all forms of inorganic nitrogen during the period of observation did not exceed the corresponding MACs (Table 10.6), which made it possible to accept them as a threshold value for AC assessment of marine ecosystems.

Analysis of the content and distribution of all inorganic forms of nitrogen showed that the priority form of nitrogen as a contaminant for all parts of the

Table 10.6 Characteristics of the content of inorganic forms of nitrogen in some areas of the water area of the Sevastopol Bay [38]

Feature	District S	W District	E-District	District C	Threshold content (MPC) (μM/L)
Number of Definitions	714	1117	750	1563	
Contents average/maximum, μM/L:					
NO_2	0.23/ 1.48	0.12/ 0.42	0.20/0.96	0.13/ 1.93	1.43
NO_3	12.6/ 142.8	2.46/ 13.3	3.25/43.0	1.71/ 13.9	221.4
NH_4	0.95/ 8.17	0.57/ 8.18	1.05/5.12	0.71/ 8.31	20.7

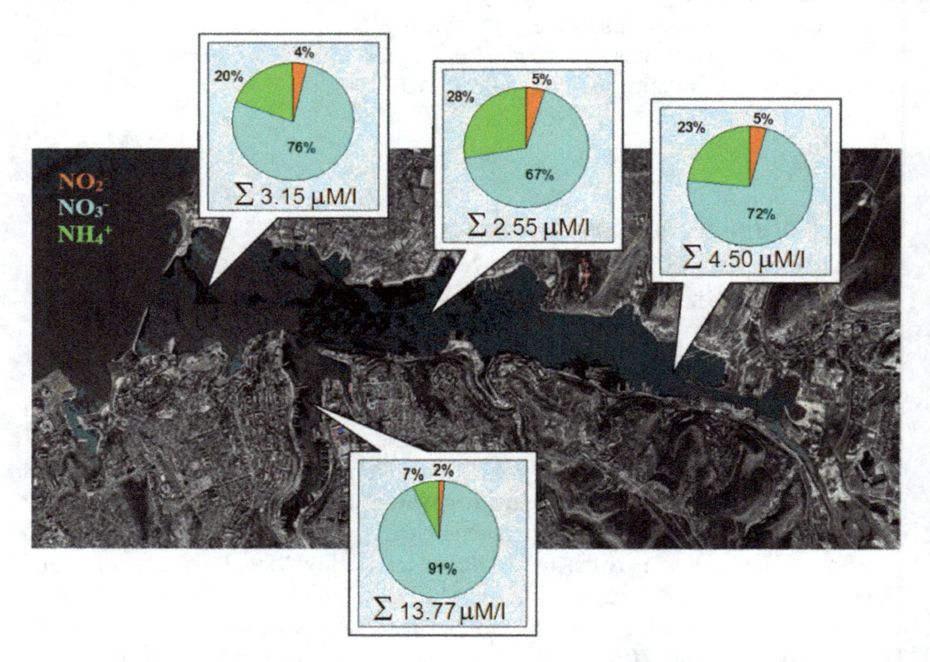

Fig. 10.12 Distribution of the average proportional content of forms of mineral nitrogen in sea waters of the selected areas of the Sevastopol Bay in 1998–2012

Sevastopol Bay were determined nitrates (Fig. 10.12), the share of which in total mineral nitrogen content is from 67% (central part of the bay) to 91% (South Bay).

In terms of total inorganic nitrogen content, on average, for the period of observations the waters of the central region were one and a half times cleaner than the adjacent waters and more than 5 times cleaner than the extremely polluted waters of the b. South Bay.

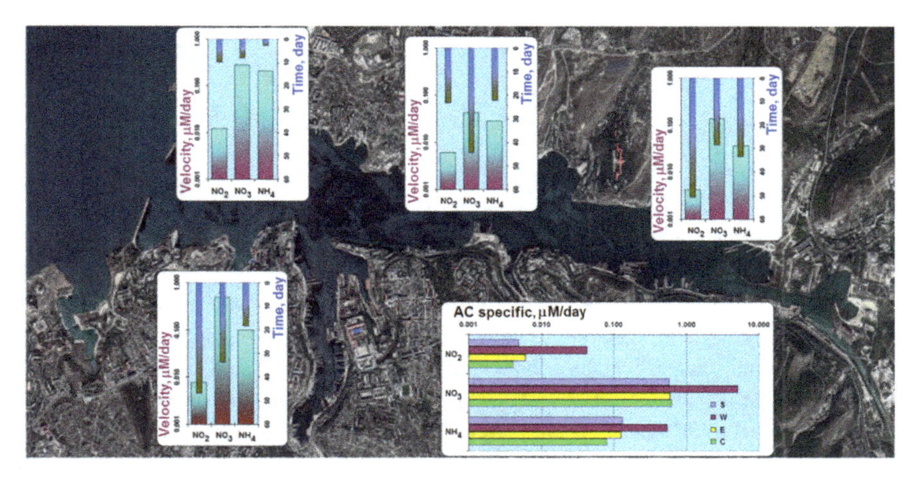

Fig. 10.13 Distribution of the rate and time of removal of inorganic nitrogen and assimilation capacity in various regions of the Sevastopol Bay in 1998–2012

The calculated average daily removal rate of mineral complexes ranged from 0.004 μM/L for nitrites in the eastern part of the Sevastopol Bay to 0.49 μM/L for nitrates in b. Southern b (Fig. 10.13). And while average daily rates for the southern part of the bay (mainly due to the nitrate form) and the western part reached 0.20 and 0.18 μM/L respectively, for the eastern part of the bay the average daily rate of mineral nitrogen removal was only 0.06 μM/L, and for the central part it was less than 0.03 μM/L.

Thus, in case of heterogeneous inflow of biogenic elements, the danger to the ecosystem is posed by seasons in which the load exceeds the specific (calculated on a unit of fixed volume, in our case by 1 L) AC value, which is in the western part of the Sevastopol bay for nitrates 5.10; nitrites 0.043; ammonium 0.54 μm per day, in the central and eastern parts of the bay an order of magnitude smaller: for nitrates 0.61 and 0.58; nitrites 0.004 and 0.006; ammonium 0.08 and 0.13 μm per day respectively.

The specific AC value of the ecosystem of the Southern Bay can be estimated at 0.58 μM for nitrates, 0.13 μM for ammonium, and 0.005 μM for nitrites (Fig. 10.13).

It is important to note that specific AC values obtained by calculation for each form of inorganic nitrogen differ not only from the corresponding MACs, but also differ significantly for individual water areas, allowing a more accurate assessment of the limit of self-cleaning capacity of a particular ecosystem.

Integrated assessment of water quality was carried out using the E-TRIX water trophicity index as a function of deviation from 100% oxygen saturation of water, associated with the characteristics of primary phytoplankton production (content of photosynthetic pigments, mainly chlorophyll a) and the concentration of nutrients.

According to the paper [39], the trophicity index is determined by the following formula

$$\mathrm{E-TRIX} = (\lg[\mathrm{Ch}\ \mathrm{D\%O}\ \mathrm{N} \cdot \mathrm{P}] + 1.5)/1.2$$

where Ch—concentration of chlorophyll a, μg/dm^3; D%O—deviation in absolute values of dissolved oxygen content from 100% saturation; N—concentration of dissolved form of mineral nitrogen, μg/dm^3; P—concentration of total phosphorus, μg/dm^3.

The data necessary for calculation of E-TRIX index about concentration of chlorophyll a, dissolved oxygen, mineral nitrogen, total phosphorus were calculated according to one-dimensional variant of water quality model and its eutrophication block [37]. For the Sevastopol bay the model was calibrated and proved to be good at calculation of hydrochemical characteristics of the bay as a whole and its separate parts [40]. The input parameters of the model were meteodata (wind speed and direction, air temperature, photosynthetically active radiation, humidity and cloudiness score) and data on the annual consumption and flow of dissolved substances of the rivers flowing into the water area. Annual course of transparency, values of sea water temperature, salinity, phytoplankton concentration, biogenic elements, oxygen, organic phosphorus and organic nitrogen, which are set for January 1 of the calculation year, were also used.

To estimate the level of trophicity of waters in the Sevastopol Bay, the annual course of chemical and biological characteristics of water quality used in the calculation of E-TRIX index was calculated (Fig. 10.14).

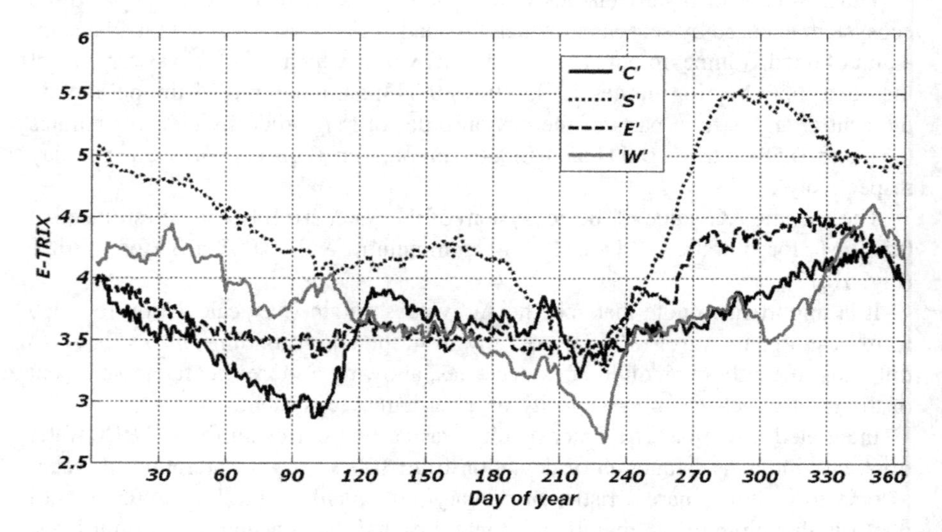

Fig. 10.14 The annual variation of the E-TRIX index of the central (C), eastern (E), western (W) regions and the Yuzhnaya Bay (S) [41]

The Southern Bay was the most polluted in almost the entire calculation year (E-TRIXmean = 4.49). The water quality of the South Bay can be described as good with average trophicity. The minimum trophicity index value (E-TRIXmin = 3.23) is observed in August and the maximum (E-TRIXmach = 5.55) in October, which coincides with the autumn phytoplankton bloom peak. The waters of other areas of the Sevastopol Bay have approximately the same trophicity index. Water quality in the central (E-TRIXmean = 3.63), eastern (E-TRIXmean = 3.81) and western (E-TRIXmean = 3.75) areas of the Sevastopol Bay is high with low eutrophication [41].

Comparison of the E-TRIX ecosystem trophicity index with biogeochemical parameters calculated from the model of water quality with the natural capacity for self-purification by AC of the ecosystem is very promising for a reliable assessment of the ecological state of the shallow ecosystem. The comparison was made with respect to nitrate nitrogen as a priority pollutant in municipal and stormwater runoffs of all parts of the Sevastopol Bay (Fig. 10.12). As shown in Fig. 10.15, the western part of the bay was the most favorable for nitrate nitrogen in terms of AC, while the central part was the cleanest in terms of trophicity index. And only Southern bay (the southern part of the Sevastopol Bay) for both indices is characterized as the most exposed to environmental risks.

Thus, the comparison of AC characteristics of different parts of the Sevastopol Bay water area and the E-TRIX trophicity index of the marine ecosystem, taking into account the level of anthropogenic load and seasonality of biological processes (warm and cold period) allows to identify adequately the water areas most vulnerable to self-cleaning and the formation of negative environmental situations flesh to catastrophic. According to the results presented in this section of the book, such water area is the area of the Southern Bay. The second most vulnerable water area is

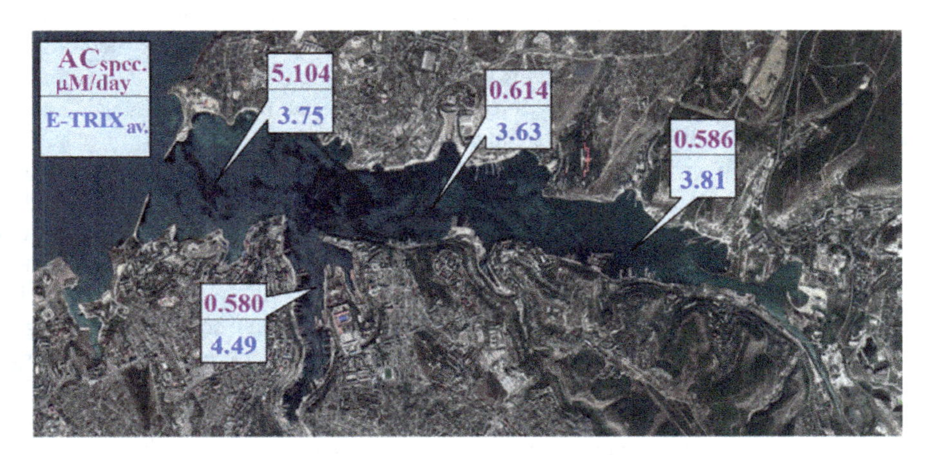

Fig. 10.15 Distribution of the AC value and the average annual E-TRIX index for nitrate nitrogen in selected areas of the Sevastopol Bay

the eastern part of the bay, which is influenced by the flow of the Chernaya River, especially during the winter-spring floods with an increase in the content of inorganic forms of nitrogen in the flow of the Chernaya River.

The observed discrepancy in the distribution of E-TRIX trophicity indices with AC in different parts of the Sevastopol Bay is due to the complexity of AC, reflecting the processes of different nature associated with both the biological cycle of biogenic elements and their chemical and dynamic removal from the ecosystem, in contrast to the E-TRIX index, which depends mainly on seasonal variability in the ecosystem of biogenic elements.

10.7 Conclusions

Analysis and evaluation of existing methods for quantitative determination of self-cleaning capacity of marine ecosystems (balance and synoptic) by means of calculation of assimilation capacity (AC) of marine shallow water ecosystems and examples of their implementation for shallow water areas of the Black Sea, which differ both in level of anthropogenic load and in hydrologic-hydrochemical and geographical characteristics, are presented. The example of individual sea areas shows the expediency of spatial (zoning) and temporal (seasonality) detailing of the assessment of the assimilation capacity of ecosystems in relation to various pollutant complexes. The necessary conditions for the application of the balance and "synoptic" methods for the calculation of the assimilation capacity are described, indicating the advantages and difficulties of each.

By the example of using two methods of AC calculation for marine ecosystems (balance and synoptic) it is shown that the synoptic method, implemented for the areas not covered by the monitoring studies, allows to conditionally assess the ecological state of the ecosystem only for the survey period. The balance method of AC assessment of marine ecosystems, taking into account their longer term functioning, allows, if necessary, to specify the AC value taking into account annual and/or seasonal changes in self-cleaning ability.

Based on the possibilities of application of the above mentioned AC calculation methods, the results of AC calculation by the balance method in relation to oil products and phenols as priority pollutants based on the materials of long-term monitoring studies are presented in this section for the ecosystems of Odessa Port and Dnieper estuary.

By the example of the Dnieper estuary for the areas of state monitoring having complex geographical and hydrologic and hydrochemical characteristics, as well as heterogeneity of anthropogenic load, in accordance with the algorithm of AC assessment implementation by the balance method, the expediency of preliminary water area zoning with subsequent assessment of self-cleaning ability for each selected area is shown.

As a result of monitoring observations in 1996–2006, AC of the Dnieper estuary ecosystems was assessed with preliminary zoning of its water area by the level of anthropogenic load and water area of the port of Odessa in relation to the priority ones for each pollutants area.

The AC of the Dnieper estuary with respect to the OP was 7960 ± 2930 tons per year. For the eastern and central regions of the estuary and the Kinbur Strait, the specific AC value was 2.17–2.31 mg/(dm^3-year), for the adjacent part of the Black Sea—2.74 mg/(dm^3-year). For the western part of the estuary it is much lower, at 1.18 mg/(dm^3-year).

For the water area of Odessa port AC with respect to phenols (amount) can be estimated at 0.042 ± 0.011 tons per year, for OP—93.5 ± 61.4 tons per year. For shallow sea areas with a significant amplitude of intra-annual temperature regime, the expediency of seasonal detailing of AC assessment is shown by the example of Odessa port water area.

At the site located in the water area of the Heracleian peninsula and subject to the indirect influence of the Sevastopol bays (Sevastopol, Krugloi, Streletskaya, Cossack, Kamyshova and Balaklava) data on the content of oil products were obtained in spring and autumn 2016.

According to the 2016 spring and autumn survey data, the synoptic method calculated the assimilation capacity of the testing area ecosystem in relation to oil products, as well as the environmental impact on the ecosystem of the investigated water area. It is shown that in spring the explored water area is characterized by ecological well-being in relation to oil products, and in autumn this water area was assessed as ecologically unfavorable as a result of the growth of recreational services.

For the Sevastopol Bay, as a water body with complex geographic and hydrologic-hydrochemical characteristics, including the presence of fresh water sources and zones of their mixing with sea water, as well as heterogeneity of anthropogenic load, it is necessary to pre-zoning the water area, in which case the ability to self-clean—the assimilation capacity of the ecosystem (AC) is calculated for each designated area. In the section of the book the self-cleaning capacity of the eastern, central, western parts of the bay and the southern bay, as well as the water area of the entire Sevastopol Bay is analyzed.

According to the data of monitoring observations of the inorganic nitrogen content in sea waters, the balance method was used to assess the AC ecosystems of the selected parts of the Sevastopol Bay in relation to nitrites, nitrates and ammonium, as priority elements of the biogenic complex in stormwater and municipal runoffs. By the example of the central part of the Sevastopol Bay the AC assessment of the marine ecosystem in relation to OP, as a priority indicator for areas of developed shipping, performed using the synoptic method, is shown.

Comparative assessments of AC values of different parts of the Sevastopol Bay water area and the E-TRIX trophicity index of the marine ecosystem, taking into account the level of anthropogenic load and seasonality of biological processes (warm and cold period) allows to adequately identify the water areas most vulnerable to self-cleaning and formation of negative environmental situations flesh to catastrophic. According to the results obtained, such water area is the ecosystem of the Southern Bay. The second most vulnerable is the ecosystem of the eastern part of the bay, which is influenced by the flow of the Chernaya River, especially during the winter-spring floods with an increase in the content of inorganic forms of nitrogen in the flow of the Chernaya River.

The observed discrepancy in the distribution of E-TRIX trophicity indices for different parts of the Sevastopol Bay with AC is due to the complexity of AC, reflecting the processes of different nature associated with both the biological cycle of nutrient elements and their chemical and dynamic removal from the ecosystem, as opposed to the E-TRIX index, which depends mainly on seasonal variability in the ecosystem of nutrient elements.

References

1. Israel YA, Tsyban AV (1989) Anthropogenic ecology of the ocean. Hydrometeoisdat, Moscow, p 528 p
2. Israel YA, Tsyban AV (1983) About the assimilation capacity of the World Ocean (in Russian). Dokl of the USSR Academy of Sciences, T. 272, № 3, pp 702–705
3. Stebbing A (1981) Assimilative capacity. Stebbing A Mar Pollut Bull 12(№ 11)
4. Israil YA (1984) Ecology and control of the natural environment: [Ezd.2-e]. Hydrometeoisdat, Moscow, 560 p
5. Israel YA, Tsyban AV, Ventzelm MV, Shigaev VV (1988) Scientific justification of the ecological normalization of the anthropogenic impact on the marine ecosystem (by the Baltic Sea example). Oceanology T. 28(№ 2), 293–299
6. Monakhov SK, Kurapov AA, Popova NV (2005) Assessment of the water area assimilation capacity and ecological rationing of pollutants discharge into the sea. Herald of NDC RAS T.20, pp 58–65
7. Sovga EE, Mezentseva IV (2008) Maintenance of oil products in sea water in the water area of Odessa port in 1997–2006. In: Ecological safety of coastal and shelf zones and integrated use of shelf resources, vol 17, pp 290–297
8. Sovga EE, Mezentseva IV, Khmara TV, Slepchuk KA (2014) About prospects and possibilities of an estimation of the Sevastopol bay water area self-cleaning ability (in Russian). In: Ecological safety of the coastal and shelf zones and complex use of the shelf resources. Sevastopol: ECOSY-Hydrophysics, vol 28, pp 153–164
9. Sovga E, Mezentseva I, Verzhevskaia L (2015) Assimilation capacity of the ecosystem of sevastopol bay. In: Ozhan E (ed) Proceedings of the twelfth international conference on the mediterranean coastal environment MEDCOAST' 2015, vol 1, Varna, Bulgaria, 6–10 Oct 2015, , pp 317–326
10. Sokolova VV, Svetasheva DR, Dzerzhinskaya IS et al (2011) Assessment of the assimilation potential and the assimilation capacity of the North Caspian Sea in relation to oil pollution. Environ Prot Oil Gas Complex:40–44
11. Ivanov VA, Katunina EV, Sovga EE (2016) Estimates of the anthropogenic impacts on the water area ecosystem of the Heracleian Peninsula in the area of the deep runoffs location №5 (1):62–68
12. Egorov VN (2001) Normalization of the anthropogenic pollution flows in the Black Sea regions by the biogeochemical criteria (in Russian). Ecol Sea № 57:75–84
13. Mezentseva IV, Klimenko NP, Khomenko ON (2009) Pollutants in the Dnieper liman water (in Russian). In: Ecological safety of the coastal and shelf zones and complex use of the shelf resources, № 18, pp 38–47

14. Yastreb VP, Khmara TV (2007) Water salinity as a condition of existence of the open estuaries ecosystems (in Russian). In: Ecological safety of the coastal and shelf zones and complex use of the shelf resources, № 15, pp 346–358
15. Sovga EE, Mezentseva IV, Lyubartseva SP (2012) Scientific substantiation of normalization of pollutant discharges in the impact areas of the Black Sea on the example of the Dnieper estuary and the water area of the port of Odessa. In: Proceedings of the international conference "Modern fishery and environmental problems of the Azov-Black Sea region", vol l, pp 233–240
16. Sovga E. Lyubartseva SP, Mezentseva IV (2010) Estimation of the Odessa port water area ecosystem capacity for self-cleaning with respect to phenols and oil products. In: Ecological safety of the coastal and shelf zones and complex use of the shelf resources, vol 22, pp 303–309
17. Sovga EE, Mezentseva IV, Lyubartseva SP (2011) Otsinka asymilianyi mysterious ecosystems of Dniprovskogo limanu schodo naftoproduktov as a method of normalizing their discount in the water area liman. Dopovidi National Academy of Sciences of Ukraine. Mathematics, Nature Studies, Technical Sciences, № 10, pp 105–109
18. Mezentseva IV (2012) Seasonal change of an assimilation capacity of the marine ecosystem in relation to the oil products on the example of the water area of the port Odessa (in Russian). Coll. of scientific articles. In: Proc. of XIV intern. "Ecological safety of the coastal and shelf zones and integrated use of the shelf resources". Exhibit. 26. Sevastopol, pp 269–274
19. Ryabinin AI, Klimenko NP, ShibACva SA (eds) (1998) Yearbook of sea water quality on hydrochemical indicators [T.1. Black Sea]. Sevastopol, MB UkrNIGMI archive
20. Sovga EE, Kondratyev SI, Godin EA, Slepchuk KA (2017) Seasonal dynamics of the content and local sources of biogenic elements in the waters of the coastal water area of the Heraclean peninsula. Mar Hydrophys J 1:56–67
21. Monakhova GA, Abdurakhmanov AM, Ahmedova GA et al (2011) Assessment of the water area assimilation capacity of the license area "North-Caspian area" in respect to hydrocarbons using a new "synoptic" method. Geogr Geoecol South Russia: Ecol Dev 4:207–212
22. Ivanov VA, Sovga EE, Katunina EV, Kotelianets EA (2017) Seasonal variability of the self-purifying ecosystem of the coastal water area of the Heraclean Peninsula with regard to oil products (in Russian). Processes in GeoMedia. 3(12):586–592
23. Ovsyanyi EI, Romanov AS, Minkovskaya RY et al (2001) The main sources of the Sevastopol region marine environment pollution. Ecological safety of the coastal and shelf zones and integrated use of the shelf resources: collection of scientific articles of NAS of Ukraine, MGI, OF InBUM. Sevastopol, vol 2, pp 138–152
24. Storkozov NA (2010) Morphometric characteristics of the Sevastopol and Balaklava Bays—environmental safety of coastal and shelf zones and integrated use of shelf resources. Sevastopol: ECOSY-Hydrophysics, Edition 23, pp 198–208
25. Ivanov VA, Ovsyanyi EI, Repetin LN, Romanov AS, Ignatyeva OG (2006) Hydrologic and hydrochemical regime of the Sevastopol bay and its changes under the influence of climatic and anthropogenic factors. Preprint. Sevastopol: MGI NAS of Ukraine, 90 p
26. Sovga EE. Mezentseva IV, Kotelianets EA (2017) Assimilation capacity of the shallow water ecosystems with different level of anthropogenic load as a method to assess their self-purification capacity. In: Problems of ecological monitoring and ecosystem modeling (PEMME), № 4, pp 39–52
27. Determination of Mass Concentration of Oil Products in Water (Methodological Instructions) of Flours 4.1.1013-01Ministry of Health of Russia, Moscow 2001. Electronic resource
28. http://www.gosthelp.ru/text/MUK41101301Opredeleniemas.html
29. Shavrak EI (2013) Assimljatsionnaja capacity of the Tsimlyansk water reservoir and stability of accumulation processes. VSU Bull Geogr Geoecol Ser 2:93–98
30. Gevorgiz NS, Kondratyev SI, Lyashenko CV, Ovsyany EI, Romanov AS (2002) Results of hydrochemical structure monitoring of Sevastopol Bay in warm period of the year. In: Ecological safety of the coastal and shelf zones and integrated use of the shelf resources. Sevastopol: ECOSY-Hydrophysics, vol 6, pp 131–148

31. Ivanov VA, Mezentseva IV, Sovga EE, Slepchuk KA, Khmara TV (2015) Assessments of the self-cleaning capacity of the Sevastopol Bay ecosystem in relation to inorganic forms of nitrogen. Processes in GeoMedia, № 2, pp 55–65
32. Sovga EE, Khmara TV (2020) Influence of the Black Sea river runoff during the flood and low water periods on the ecological condition of the Kutova part of the Sevastopol bay water area (in Russian). Mar Hydrophys J № 1:31–40
33. Ovsiany EI, Artemenko VM, Romanov AS, Orekhova NA (2007) Black river drain as a factor of the water-salt regime formation and the ecological state of the Sevastopol bay (in Russian). In: Ecological safety of the coastal and shelf zones and complex use of the shelf resources. Sevastopol, vol 15, pp 57–65
34. Badiukov DD, Korneeva GA, Savenko AV (2014) Structurally functional characteristics transformation of the mainland runoff of the Black river and the Sevastopol bay sea waters in the winter period (in Russian). Probl Reg Ecol № 3:7–13
35. Mezentseva IV, Sovga EE (2019) Self-purification ability of the ecosystem of the eastern extremity of the Sevastopol bay in relation to the inorganic forms of nitrogen (in Russian). In: Ecological safety of the coastal and shelf sea zones, vol 1, pp 71–77
36. Sovga EE, Mezentseva IV (2019) Ecological condition of the central part of the Sevastopol bay water area depending on the anthropogenic load level (in Russian). In: Ecological safety of the coastal and shelf sea zones, № 3, pp 52–60
37. Ivanov VA, Tuchkovenko YS (2006) Applied mathematical modeling of water quality in shelf marine ecosystems. Sevastopol, 368 p
38. Sovga EE, Mezentseva IV, Slepchuk KA (2020) Comparison of self-cleaning ability of ecosystems of different parts of the Sevastopol bay water area by the size of their assimilation capacity and trophicity index. In: Ecological safety of the coastal and shelf sea zones, № 3, pp 272–275
39. Vollenweider RA, Giovanardi F, Montanari G et al (1998) Characterization of the trophic conditions of marine coastal waters with special reference to the NW Adriatic Sea: proposal for a trophic scale, turbidity and generalized water quality index. Environmetrics. T. 9, vol 3, pp 329–357. https://doi.org/10.1002/(sici)1099-095x(199805/06)9:33.0.co;2-9
40. Slepchuk KA (2019) Estimation of the Sevastopol bay area eutrophication level according to the E-TRIX index numerical modeling results (in Russian). In: Processes in the geospheres, № 1(19), pp 91–96
41. Slepchuk KA, Khmara TV, Man'kovskaya EV (2017) Comparative assessment of the trophic level of the Sevastopol and Yuzhnaya Bays using E-TRIX index. Phys Oceanogr 5:67–78

Chapter 11
Environmental Monitoring, Its Main Types and Role to Solve Problems of Seas and Oceans Pollution

11.1 Introduction

At the dawn of the third millennium, humankind is facing two fundamental problems—the exploration of outer space and the oceans. And if in the field of space exploration works are intensively developed with the implementation of the latest technologies, in the field of study of the World Ocean research is carried out rather less intensively and is mainly related to the search for resource bases and support of fishing activities [1].

The term "monitoring" first appeared in the recommendations of the Special Commission of SCOPE (Scientific Commission on Problems of the Environment) of UNESCO in 1971.

The need to identify non-native changes in the structure and functioning of marine ecosystems, and to normalize human impacts on the ocean, has created a need for integrated global monitoring of the ocean. This monitoring is being developed as part of the Global Environment Monitoring System Operation Center (GEMSOC) program, a work center for the programme was established in 1975 in Nairobi, Kenya). In the USSR, and then in Russia, work is being carried out under the MONOC programme. The programme was developed at the Institute of Global Climate and Ecology (IGCE) of Roshydromet and the Russian Academy of Sciences (until 1991, Laboratory for Environmental and Climate Monitoring). The MONOC programme was developed taking into account the recommendations of the I International Symposium "Integrated Global Monitoring of the World Ocean", which was held in Tallinn, October 2-10, 1983.

Environmental monitoring of the ocean, as a system of observations, analysis, assessment and prediction of the state of the ocean, includes physical, geochemical and biological components. The physical component of the environmental monitoring of the ocean includes the systematic analysis, observation and prediction of thermodynamic and proliferation processes of anthropogenic pollution that determine the environmental situation in the ocean. The physical component of

K. Pokazeev et al., *Pollution in the Black Sea*, Springer Oceanography,
https://doi.org/10.1007/978-3-030-61895-7_11

environmental monitoring is closely linked to climate monitoring and therefore research in this field is part of the World Climate programme. Much of the programme's attention is devoted to studying the effects of ocean properties and dynamics on heat and gas exchange with the atmosphere, on the global cycle of heat, moisture and various chemical compounds, especially carbon dioxide, in the climate system. It is also tasked with determining the impact of various ocean-atmosphere processes on the climate system, including the cryosphere.

The geochemical component of environmental monitoring of the ocean includes systematic observation, assessment and prediction of levels of pollution of marine ecosystems, including the rate of entry of pollutants into the world's oceans, their content in seawater, accumulation in suspended matter, bottom sediments, biota; rates of removal of pollutants from seawater through nutrient sedimentation and microbial metabolism.

The biological component of environmental monitoring of the ocean includes systematic observation, assessment and prediction of biological effects of anthropogenic pollution and other negative impacts on marine ecosystems. Its tasks include identifying "critical" impact factors and the most vulnerable links in the biotic component of marine ecosystems.

The launch of global monitoring systems is a qualitatively new stage in the development of monitoring and it is very expensive. Global systems are being developed through major international programmes, such as NASA's Ocean Processes Program, which aims to understand the role of the oceans in shaping the Earth's climate. The experience of this programme shows the reality of global monitoring systems that accumulate data such as surface temperature distribution, wind vectors over water, radiation balance, gas exchange at the border with the atmosphere, etc. These data can be used to create models to forecast the state of ocean ecosystems.

The world's largest fuel and energy companies involved in the production of hydrocarbons on the sea shelf are systematically working to improve the environmental safety of marine operations, including the improvement of the environmental monitoring. Environmental monitoring of sea areas in the course of production activity promotes convergence of interests of the state and business related to the use and development of the shelf zone.

Organization of effective environmental monitoring of industrial activity of fuel and energy complex enterprises on the sea shelf becomes more and more urgent, first of all, due to expansion of hydrocarbon production and transportation, which in emergency situations can lead to negative consequences for the coastal territories. According to the Marine Doctrine of the Russian Federation for the period until 2020, approved by Presidential Decree №PR-1387 of 27.07.2001, "prevention of marine pollution" is one of the main provisions relating to the safeguarding of national interests in the World Ocean. One of the principles of the national marine policy is "development of systems for monitoring the state of marine natural environment and coastal areas".

Particularly relevant are the tasks of environmental monitoring in the areas of oil and gas deposits in the Caspian Sea (see "Special environmental and fishery requirements for geological study, exploration and production of hydrocarbons in the reserve zone in the northern part of the Caspian Sea in the license areas "North", "Vostochno-Rakushechnaya" and "North-Caspian area", approved by the Ministry of Natural Resources in 2005). In recent years, much more attention has been paid to the environmental monitoring of marine areas than before. In particular, the P. P. Shirshov Institute of Oceanology of the Russian Academy of Sciences (IORAN) has performed works within the framework of the project of the Ministry of Education and Science (RP-22.1/001) "Creation of a system of multilevel regionally-adapted environmental and geodynamic monitoring of the seas of the Russian Federation, primarily the shelf and continental slope", which resulted in the development of a system of information support for industrial and environmental safety of objects of oil and gas field development on the sea shelf [2].

A network of continuous monitoring of subsoil use facilities in the marine environment should provide information on key parameters of the marine environment in real time in order to assess current impacts on subsoil use facilities on the one hand, and critical components of marine environmental systems on the other. The main task of operational environmental monitoring is to control possible anthropogenic pollution. Early detection of contaminants leaks is necessary for the timely adoption of measures to prevent major accidents. Long-term monitoring data are used to monitor the status of key environmental parameters and to identify anthropogenic factors against the background of natural trends. This allows predicting negative consequences and making adequate decisions to minimize the risk of damage.

The main objectives of environmental monitoring of the ocean are to establish an observing system for sources and factors of anthropogenic impacts and biological effects in marine ecosystems, and to determine the maximum allowable load on ecosystems (developed on the basis of assessment, analysis and prediction of the state of the ocean).

The environmental monitoring system accumulates, systematizes and analyzes information:

- about the state of the environment;
- about the reasons for the observed changes in condition;
- about acceptability of changes and stresses on the environment as a whole;
- about existing biosphere reserves.

11.2 Types of Environmental Monitoring Their Purpose and Objectives

Environmental monitoring is understood as a set of activities to determine the degree of contamination of ecosystems or elements of the biosphere, observation of imbalances in the ecological balance. A distinction is made between the following types of monitoring: **global, national, regional, local**.

There is a global ocean observing system. Russia is a member country of this programme. The Unified State Information System on the World Ocean (ESIMO) has been developed. The GIS-server of ESIMO provides more than 150 layers with observed, diagnostic and prognostic hydrometeorological and ice information about the world ocean and individual seas of Russia, as well as the adjacent land. The data are promptly updated at intervals of 3 h or more. The interactive map contains conditionally constant spatial data (countries, administrative areas of Russia, cities, rivers and canals, etc.), as well as information about Russian ports.

The global environmental monitoring system consists of five interrelated items:

- a study of climate change;
- long-range transport of pollutants;
- hygienic aspects of the environment;
- the study of the world's oceans;
- a study of sushi resources.

Currently, there are 22 networks of active global monitoring stations, as well as national and international monitoring systems. One of the main ideas of monitoring is to reach a fundamentally new level of competence in decision-making on local, regional and global scales.

The efficiency of any monitoring system (type) is largely determined by its organization, which is a complex, multifaceted task.

First of all, the complexity of monitoring organization depends on its level. Environmental monitoring can cover local territories (district, region)—the local level, individual regions (districts)—the regional level, and the globe as a whole—the global level. In this case, taking into account the level of monitoring, a sufficient network of stations, points, observation posts, equipped with the most modern equipment, using the latest technologies should be developed.

The ratio of environmental monitoring levels is shown in Fig. 11.1.

11.2.1 Global Monitoring

Global monitoring is the monitoring of world processes and phenomena in the Earth's biosphere and its ecosphere, including all their ecological components (the main material and energy components of ecological systems) and warning of emerging extreme situations. Under the increasing pressure of industrial civilization, pollution is becoming a global factor determining the development of the natural environment and human health.

Creation of global monitoring systems is a qualitatively new stage in the development of monitoring and is very expensive. Global systems are being developed through major international programmes, such as NASA's Ocean Processes Program, which aims to understand the role of the oceans in shaping the Earth's climate. The experience of this programme shows the reality of global

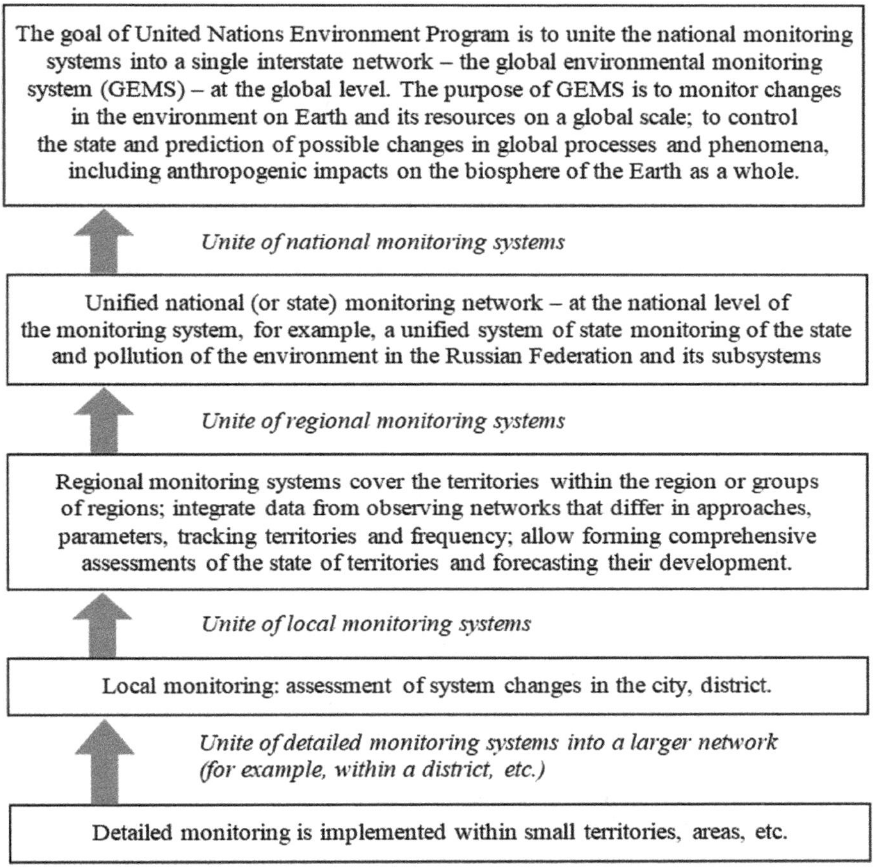

The goal of United Nations Environment Program is to unite the national monitoring systems into a single interstate network – the global environmental monitoring system (GEMS) – at the global level. The purpose of GEMS is to monitor changes in the environment on Earth and its resources on a global scale; to control the state and prediction of possible changes in global processes and phenomena, including anthropogenic impacts on the biosphere of the Earth as a whole.

Unite of national monitoring systems

Unified national (or state) monitoring network – at the national level of the monitoring system, for example, a unified system of state monitoring of the state and pollution of the environment in the Russian Federation and its subsystems

Unite of regional monitoring systems

Regional monitoring systems cover the territories within the region or groups of regions; integrate data from observing networks that differ in approaches, parameters, tracking territories and frequency; allow forming comprehensive assessments of the state of territories and forecasting their development.

Unite of local monitoring systems

Local monitoring: assessment of system changes in the city, district.

Unite of detailed monitoring systems into a larger network (for example, within a district, etc.)

Detailed monitoring is implemented within small territories, areas, etc.

Fig. 11.1 Correlation of environmental monitoring levels

monitoring systems that accumulate data such as surface temperature distribution, wind vectors over water, radiation balance, gas exchange at the border with the atmosphere, etc. These data can be used to create models to forecast the state of ocean ecosystems.

The objective of global monitoring is to provide observations of control and prediction of changes in the biosphere as a whole, so it is also called biosphere or background monitoring. The Global Environmental Monitoring System (GEMS) is developed and coordinated by UNEP and the World Meteorological Organization through various international programmes and projects. The Global Environmental Monitoring System is shown in a diagram (Fig. 11.2).

The need for integrated global monitoring of the ocean has arisen in response to the need to identify un-natural changes in the structure and functioning of marine ecosystems, and to normalize human impacts on the ocean.

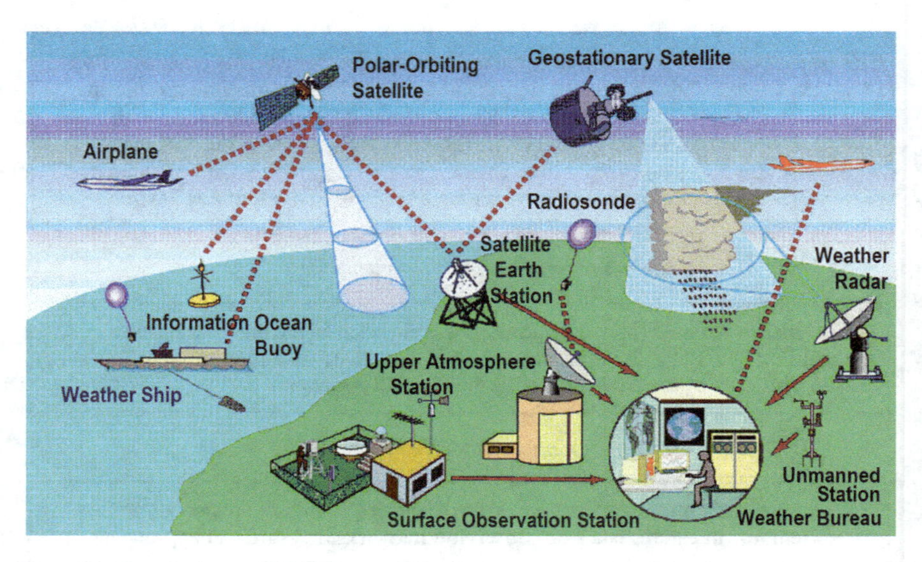

Fig. 11.2 Global observing system

The paper [1] discusses the concept of development and establishment of the Global System for Environmental Monitoring of Seas and Oceans. The system can be created as a set of subunits in the form of international and national subsystems of corresponding waters using the same principles of structural organization, including communication channels and information transfer and processing points. Each subsystem includes the following mandatory structural elements, the number of which varies depending on the size of the served water area: nuclear submarines or diesel-electric submarines in the form of converted submarine research vessels capable of operating with a wide range of underwater unmanned vehicles; sea-planes; research vessels equipped with unmanned aerial vehicles; space communication system; space surveillance and control systems; scientific information center The system is created as a set of closed and open joint stock companies integrated into the corresponding Consortium.

Achieving environmental monitoring of the world's oceans requires a number of special tasks.

1. Identification of inflow channels and assessment of pollutant flows in bio-productive and lightly curbed ecosystems of the world's oceans.
   The solution of this problem is based on field observations, that allow to identify the main sources and channels of pollutants withdrawal, to evaluate the processes of self-cleaning of the marine environment, to calculate balances of pollutants in certain regions of the ocean, to describe the dynamics of toxic substances in the components of marine ecosystems and to study their biochemical cycles. Of particular practical importance is the study of pollutant inputs and destruction in the most productive areas of the ocean, in the PMS and in the water column.

2. Study of the negative effects of pollution of bio-productive and lightly protected ecosystems in the world's oceans.
   Today's understanding of the environmental impacts of ocean pollution is still being developed mainly through coastal studies. Vast areas of the world's oceans are barely covered by such studies. Therefore, long-term observations of the state of the neuston, plankton and benthic communities, whose structure is subject to cyclic oscillations under the influence of different natural phenomena, must be carried out to obtain the necessary information. The task is to identify those changes that are determined by anthropogenic factors against the background of natural fluctuations in the properties of marine ecosystems.
3. Study of causal relations between levels of accumulation of pollutants and observed environmental changes. Determination of critical concentrations of pollutants that can cause disruptions in biological processes.
   At present, there is a clear lack of information on the causal links between contaminant concentrations and disturbance in marine ecosystems. The search for this information is driven by the need to identify critical concentrations of contaminants that, if slightly exceeded, could lead to a necessary reduction in the resilience of the entire ecosystem if affected by a species or group of organisms.
4. Study of the physical, chemical and biological processes that determine assimilation capacity and assessment of the assimilation capacity of marine ecosystems in the most studied areas of the world's oceans.
   Marine ecosystems have a wide range of physical, chemical and biological mechanisms by which pollutants can be removed from it without serious disturbance of the biological cycle. But when concentrations of pollutants in the environment reach levels that exceed the ecosystem's assimilation capacity, they begin to affect the survival, growth and reproduction of hydrobionts. Determining the assimilation capacity is necessary to normalize external influences.
5. Mathematical models of selected environmental processes to forecast the ecological situation in the ocean at local, regional and global scales.
   Predicting changes in natural ecosystems, their composition, structure and resilience is one of the most important tasks of marine environmental research. Such predictions are possible through mathematical modelling of ecosystem behavior, taking into account internal and external relationships. The development of such models is an important and responsible task for scientists.

11.2.2 Regional Environmental Monitoring

Regional monitoring—covers individual regions in which processes and phenomena are observed that differ from natural or anthropogenic impacts. In other words, regional monitoring tracks processes and phenomena within a region where these processes and phenomena may differ in nature and anthropogenic impacts from the background (baseline) characteristic of the entire biosphere.

Regional monitoring is organized in the territories of large regions of large states, for example, such as: Russian Federation, USA, Canada, etc. Regional monitoring is not only a part of national monitoring, but also solves tasks specific to the territory. In addition, regional monitoring can be international if a region includes the sea (e.g. Caspian, Black or Baltic Sea) or other natural formations that require special attention from several countries, such as the Great Lakes on the US-Canada border.

The subject of regional monitoring, as its very name implies, is the state of the environment within a region.

Regional monitoring involves organizations of ministries and agencies responsible for national monitoring and local environmental organizations. An important subsystem of regional monitoring is the monitoring of sources of pollutants, based on which the list of pollutants to be determined in different natural environments is specified, and the contribution of global and regional sources of pollutants to changes in the region's natural environment is determined.

The results of the regional monitoring are used by local governments responsible for environmental activities and decision making in this area. The network of regional monitoring stations is organized taking into account physical and geographical conditions, location of industrial, energy and agricultural enterprises, population distribution in the region.

The main objective of regional monitoring is to obtain more complete and detailed information on the state of the region's environment and its impact on human beings, which cannot be done in the framework of global and national monitoring, as their programmes cannot take into account the specific features of each region.

In the area of regional monitoring in Russia, observations are conducted mainly by Roshydromet, which has an extensive network, as well as by some agencies (agrochemical service of the Ministry of Agriculture, water and sewage service, etc.) And finally, there is a network of background monitoring, carried out under the MAB (Man and Biosphere) program. Small towns and numerous settlements, the vast majority of diffuse sources of pollution, remain practically uncovered by the monitoring network. Monitoring of the aquatic environment, organized primarily by Roshydromet and to some extent by Sanitary and Epidemiological (SES) and Communal (Vodokanal) services, does not cover the vast majority of small rivers. At the same time, it is known that pollution of large rivers is largely due to the contribution of their extensive tributaries and economic activities to the catchment. With the reduction in the total number of observation posts, it is clear that the state currently does not have the resources to organize any effective monitoring system for small rivers.

Economic monitoring—a system of observations at the regional level on environmental changes in the process of nature management, especially in intensively developed areas.

The regional level of the unified state system of environmental monitoring (USSEM) is designed to solve the tasks of environmental monitoring, which are of regional nature, with the definition of territorial subsystems of USSEM, participating in the formation of the regional system.

The expediency of creating a regional level of ESSEM is determined:

- The need to assess the state of natural objects, analysis of natural processes and environmentally unfavorable phenomena, when their borders do not coincide with the borders of the subjects of the Russian Federation;
- The existing structure of territorial (regional) bodies of a number of agencies;
- The expediency of creating powerful regional functional center capable of serving a number of constituent entities of the Russian Federation.

The most important component in the organization of the ESSEM is the organization of monitoring of sources of anthropogenic impact on the environment (local monitoring).

11.2.3 Local Monitoring

Local environmental monitoring is a system of continuous observation of impact of a specific economic and other activities on environment. The main problems of local monitoring organization are related to the solution of three main tasks: (1) creation of a network of observation points; (2) ensuring operational control of water bodies; (3) selection of controlled parameters and indicators of water bodies condition and individual analytical parameters.

An example of local monitoring is a permanent system of monitoring and control of air pollution in cities, on transport highways, carried out by means of stationary, mobile or under flare posts. Such a system exists in most major Russian cities. Each ministry and department, in accordance with its own specifics and available regulatory documents, organizes departmental control services at its enterprises and reports to regulatory authorities of the Ministry of Natural Resources (MNR) of Russia on the adopted forms of statistical reporting and other developed and approved documents.

Organizational and production control of industrial emissions and discharges does not meet the requirements for automated measuring and information systems, the bulk of data on the quantitative composition of emissions into the atmosphere is obtained using mainly instrumental and laboratory methods of control.

Territorial subdivisions of the Russian Ministry of Natural Resources monitor sources of anthropogenic impact on the atmospheric air. The task of observations of this type is to determine the composition and quantity of pollutants coming into the environment from emission sources. Observation objects are enterprises (industrial, agricultural, transport, etc.) that have sources of environmental pollution as well as the sources themselves. These observations include:

- Departmental control, which is performed by departmental laboratories for control of industrial emissions, which have passed metrological certification, or with the involvement of licensed (accredited) laboratories;
- State control performed by industrial emission control laboratories under the territorial bodies of the Russian Ministry of Natural Resources.

Specialized Analytical Control Inspections (SACI) supervise the activities of departmental services and laboratories that determine the content of pollutants in the emissions and discharges of enterprises.

There are two types of local monitoring—facility specific industrial monitoring and impact specific monitoring.

According to the legislation, during the construction of facilities, their operation and in the post-operational period it is necessary to perform industrial environmental monitoring (PEM). The purpose of industrial environmental monitoring is to control the environmental state of the environment in the area of influence of construction and operation of the facility by collecting measurement data, their comprehensive processing and analysis, to assess the situation and make managerial decisions.

According to the degree of readiness for local environmental monitoring, all enterprises can be conditionally divided into 3 groups:

1. There are a small number of large enterprises that perform organized emission of harmful substances into the atmosphere (through chimneys). These enterprises partly have equipment necessary for monitoring, but it is domestic and does not cover the entire range of substances subject to mandatory definition. But with them the problem is solved easier: the analyzers are mounted in the chimney from which the emission takes place. These enterprises can easily be retrofitted with the necessary devices.
2. The second group are enterprises that have an organized but distributed emission, the so-called “cluster pipes”, i.e. the emission is distributed between a certain number of pipes (e.g. 50 pipes). In this case analyzers are supposed to cut into one of them.
3. The third group includes enterprises that emit unorganized emissions, i.e., those without pipes, e.g., gas depots, petrol stations, etc.

Consistent implementation at industrial enterprises of a systematic approach to environmental protection is provided in accordance with the requirements of international standards series ISO 14000, the national standard GOST and ISO 14001-2007 and Regulations № 1836-93 of 29.06.93 on the voluntary participation of industrial sector companies in the scheme of environmental management and auditing of the European Community. Within the framework of these standards, monitoring and control are understood respectively as ***industrial environmental monitoring*** (condition of natural environment in the area of organization’s location) and ***industrial environmental control*** (assessment of organization’s compliance with the requirements of environmental legislation, control of emissions (dumping) of harmful substances and waste generation).

ISO 14000 (ISO 14000), the standards for environmental management systems, offers a simple, harmonized approach to environmental management applicable to all organizations around the world. It is compatible with ISO 9000 series and DNV OHSMS (based on BS 8800). ISO 14001:2004 (ISO 14001) contains all elements of a standard management system such as strategy, goals and objectives, management programme, operational control, monitoring and evaluation, training, internal audits and management analysis. The ISO 14000 series of standards (ISO 14000) is officially voluntary. They do not replace legislative requirements, but ensure the management system of the enterprise's environmental impact and compliance with the requirements of environmental legislation. About 65,000 companies in the world currently voluntarily participate in the implementation of ISO 14000 series standards.

China, Italy, Great Britain, Japan, Spain and the United States of America are among the top ten countries in the field of environmental management system certification. Let us note the advantages of supporting the standards under consideration by enterprises:

1. ISO 14001:2004 certification is becoming increasingly clear to promote products and services on international markets, as well as to improve the company's relationship with consumers, government and local communities.
2. Increase in the estimated value of fixed assets of the enterprise.
3. Growth of competitiveness through optimal use of energy and water resources, careful selection of raw materials and controlled waste processing.
4. Decrease in financial expenses for payment of fines for violation of environmental legislation requirements.
5. Ensuring the reduction of negative environmental impact in a cost-effective manner, thus combining economic and environmental objectives.
6. Increasing the adaptive capacity of Russian companies and strengthening their market positions.

An important component of local monitoring is an inventory of sources of environmental impact. It consists of a documented description (including additional measurements) of the total number, location and main characteristics of the sources of impact, including their compliance with the established norms and limits. In the Russian Federation, a periodic, every five years, inventory of sources of pollutant emissions is mandatory.

Another important component of local monitoring is waste inventory. Waste inventory is defined as a documented description (including through supplementary measurements) of the total quantity and main characteristics of individual waste streams, as well as their disposal and disposal methods.

At present, there are many organizations provide local monitoring services (e.g. the largest project institute of the Department of the FEB of the All-Russian Research Institute of Nature). It should be noted that for large companies local monitoring often overlaps with regional monitoring. Figure 11.3 presents a scheme of responsibility distribution for various types of environmental monitoring among governmental agencies in the Russian Federation [3].

11.2.4 Impact Monitoring

Impact monitoring is carried out in especially hazardous areas directly adjacent to the sources of pollutants. In other words, impact monitoring is the monitoring of regional and local anthropogenic impacts in especially hazardous zones and places. Ideally, ***impact*** monitoring system should accumulate and analyze detailed information on specific sources of pollution and their impact on the environment.

The ecological map (Fig. 11.3) clearly indicates white spots where no systematic observations are made. Moreover, within the framework of the state environmental monitoring network, there are no prerequisites for their organization in these places. These white spots that can (and often should) become objects of public environmental monitoring. Practical orientation of the monitoring, concentration of efforts on local problems in combination with a well-thought-out scheme and correct interpretation of the obtained data allow for effective use of resources available to the public. Besides, these features of public monitoring create serious preconditions for organizing a constructive dialogue aimed at consolidating the efforts of all participants [3].

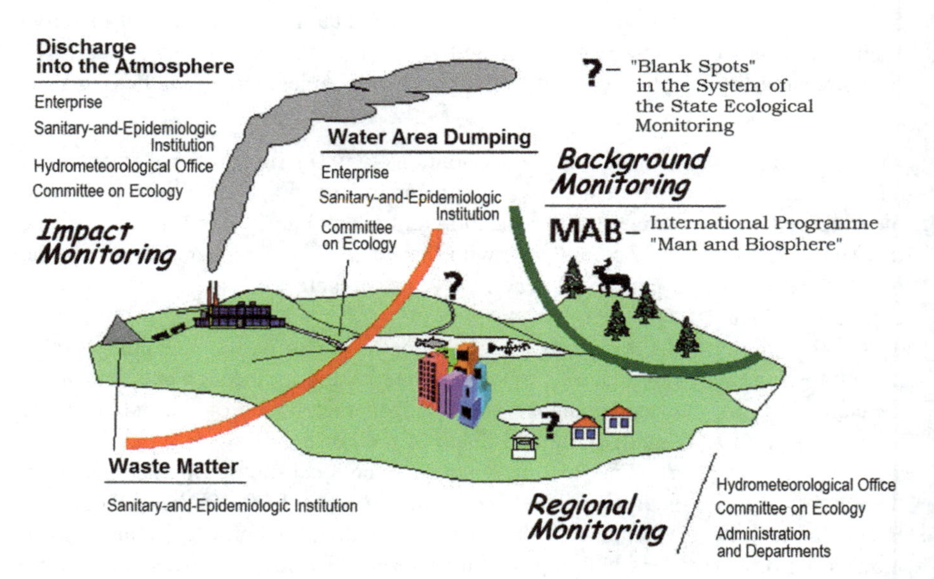

Fig. 11.3 Levels of environmental monitoring and distribution of responsibility for its implementation among government agencies in Russia

Monitoring of anthropogenic impact sources (MAIS) is aimed at solving the problem of specific (specific) impact made by a business entity on the components of the environment, and is the information basis for the development of a strategy to manage the anthropogenic impact and take appropriate management decisions.

Functioning of MAIS provides:

- making observations in the area where the enterprises are located;
- obtaining reliable information about the sources of emissions and their impact on the environment;
- information support for development of environmental measures and assessment of their effectiveness;
- improvement of ecological situation and population health in the area of anthropogenic impact sources of the economic entity.

Reliability of the information received within the framework of implementation of MAIS can be obtained by observing the unified methodological and metrological requirements, using unified formats of processing and transfer of results. Industrial enterprises are obliged to keep state environmental statistical reporting, which includes:

1. Form 2-tp (air) “Data on protection of atmospheric air”. Report on protection of atmospheric air.
   It presents annually and includes data on emissions of pollutants into the atmosphere, their purification and utilization; data on emissions of specific pollutants into the atmosphere; sources of emissions of pollutants into the atmosphere; implementation of measures to reduce emissions.
2. Form 2-tp (waterworks) “Data on water use”. Report on the use of water. It is filed annually and includes data on water taken from natural sources, received from other enterprises, used and transferred water; data on water discharge, water recycling and resupply systems; established water withdrawal limits.
3. Form 2-tp (waste) “Data on production and consumption waste generation, neutralization, transportation and disposal”. Report on formation and disposal of waste. It is filed annually and includes data on waste (availability, formation, receipt from other enterprises, use, neutralization, organized and unorganized storage and disposal) of five hazard classes of waste.
4. Form 1—sewerage. Report on the work of the sewerage. Includes data on the availability of sewerage facilities and their operation for the year.
5. Form 1—water supply. Report on the work of the water supply system. Includes data on the availability of water supply facilities and their operation for the year.
6. Form N 4—environmental payments (EP). Report on current expenditures on nature protection and environmental payments.

Ideally, an impact monitoring system should collect and process detailed data on specific sources of pollution and their impact on the environment. However, in the system established in the Russian Federation, information on the activities of enterprises and the state of the environment in the area of their impact is mostly averaged or based on statements of the enterprises themselves. Most of the available materials reflect the nature of dispersion of pollutants in the air and water, established by model calculations, and the results of measurements (quarterly—on water, annual or more rare—by air). The state of the environment is described quite fully only in large cities and industrial areas.

11.3 Conclusions

1. The main tasks of ecological monitoring of environment in the XXI century and problems arising at their fulfillment are covered.
2. The main components of environmental monitoring of the world's oceans (physical, geochemical and biological) and objectives are considered.
3. The ratios of the main types of environmental monitoring are given. The specifics, features and tasks of certain types of monitoring: global, regional, local and impact are described.
4. The distribution of responsibility for the implementation of types of environmental monitoring among government agencies in Russia is given.
5. The general difficulties arising in the implementation of regional and local environmental monitoring are considered.

References

1. Suponitskiy VL (2015) Global system of ecological monitoring of seas and oceans. Univ Bull (State University of Management) 12:134–139
2. Lobkovsky LI, Flint MV, Kopelevich OV, Zatsepin AG, Kovachev SA (2005) Technology of multilevel regionally-adapted ecological and geodynamic monitoring of the seas of the Russian Federation, first of all of the shelf and continental slope. Institute of Oceanology. In: IX international scientific and technical conference "Modern methods and means of oceanological research". Proceedings of the conference, pp 56–58
3. Strelnikov VV, Melchenko AI (2012) Ecological monitoring textbook. Krasnodar: Publishing House—South, 372 p
4. https://studopedia.ru/1_63892_tseli-i-zadachi-monitoringa-mirovogo-okeana.html
5. https://studopedia.ru/20_39775_ekologicheskiy-monitoring-okeana-ego-osnovnie-sostavlyayushchie-fizicheskaya-geohimicheskaya-biologicheskaya.htmlю

Chapter 12
Organize Environmental Monitoring of Specific Sea Areas Using the Example of the Black Sea

12.1 Introduction

The Black Sea as a part of the Azov-Black Sea region is one of the most developed regions in terms of recreational and tourist, health-resort and balneological facilities not only for Russia, but for Europe as a whole. This is primarily due to the presence of an extensive network of sources of mineral water and therapeutic mud, which in combination with unique climatic and natural recreational factors gives rise to a developed system of recreational and spa complexes [1].

At the same time, the catastrophic pollution of the Black and Azov Seas is a generally recognized fact. However, due to the lack of funds, environmental monitoring is not sufficient to adequately assess the current environmental condition of the water area. Another factor hindering monitoring activities is the lack of international legislation on the economic zone (6 independent states). At the same time, the environmental situation of the Black Sea and the Sea of Azov is deteriorating, mainly due to the intensification of economic activities, also due to the increasing concentration of industry and agriculture in the coastal zone. Therefore, monitoring data of appropriate quality serve as the basis for the preparation of a legislative framework and control of pollution as well as marine environmental management.

Data on the inflow of pollutants into the marine environment from land-based sources are needed to assess the status of high seas and coastal waters, as well as to formulate environmental policies and assess the efficiency of measures to reduce pollution from watersheds. The assessment of coastal inputs to the sea (load) takes into account three potential sources—loads from controlled rivers, uncontrolled areas and point sources that discharge waste water directly into the sea.

In the fields of activity for all seas, the main sources of pollution are municipal facilities (municipal wastewater treatment plants), commercial, oil and fishing fleets, industrial enterprises of various forms of ownership, as well as river runoff, accumulating pollutants from all point and diffuse sources in the catchment area.

K. Pokazeev et al., *Pollution in the Black Sea*, Springer Oceanography,
https://doi.org/10.1007/978-3-030-61895-7_12

In this section, the problems of environmental monitoring of the Black Sea will be considered in the following order:

- Peculiarities, structure and objectives of the Black Sea environmental monitoring means and methods adopted in the Russian Federation;
- The peculiarities, structure and objectives of the environmental monitoring of the Black Sea have been adopted in the Black Sea countries (Ukraine);
- Proposals to improve the system of environmental monitoring of the Black Sea under international projects and programes;
- Peculiarities, structure and objectives of the Black Sea environmental monitoring means and methods adopted in the Russian Federation.

In Russia, at the federal level, the Ministry of Natural Resources and Ecology of the Russian Federation (Ministry of Natural Resources and Ecology of the Russian Federation, www.mnr.gov.ru) has the responsibility to monitor the natural environment and its pollution. With regard to the implementation of state monitoring of water bodies, the Ministry of Natural Resources of Russia establishes requirements for observations of the state of the natural environment and its pollution, collection, processing, storage and dissemination of information on the state of the natural environment and its pollution, as well as to obtain information products. The Ministry of Natural Resources and Ecology of the Russian Federation coordinates and controls the activities of the Federal Service for Hydrometeorology and Environmental Monitoring (Roshydromet), the Federal Service for Supervision of Natural Resources, the Federal Agency for Water Resources and the Federal Agency for Subsoil Use under its jurisdiction. In accordance with Government Decree of the Russian Federation of 06.06.2013 № 477 “On the implementation of state monitoring of the condition and pollution of the environment” and the attached “Regulations on state monitoring of the condition and pollution of the environment” Roshydromet provides for the formation and operation of the state surveillance network, including the organization and termination of stationary and mobile observation points, including ship expeditionary research, determination of their location and the implementation of the Federal Service for Hydrometeorology and Environmental Monitoring (Roshydromet). All primary information on the results of water and pollution monitoring is sent to Roshydromet institutes, as well as to the Unified State Data Fund (USDF), Rosvodresursy and the Russian Ministry of Natural Resources for storage, processing and preparation of information products.

At present, the state observation network is formed on the basis of the State Observation Service (SOS) (2003) and includes regional Hydrometeorology and Environmental Monitoring Departments (HMMDs), while the practical monitoring work is carried out by their branches—Centers for Hydrometeorology and Environmental Monitoring (CHEM), (http://www.meteorf.ru/about/structure/). The results of the work of Roshydromet’s marine network are presented in succinct form in the “Yearbooks of sea water quality on hydrochemical indicators”. In addition, the Yearbooks include, to the extent possible, the results of research and observations of other organizations and research institutes of RosHydroMet and the

Russian Academy of Sciences, data from international information exchange, inter-resources, as well as materials from individual expeditionary marine research of state and non-governmental organizations [2].

12.2 Russian Sea Monitoring Stations of Roshydromet

Observations of the state of the marine environment in the coastal areas of the Russian seas are regularly conducted at intentionally selected sea points—marine stations. Stations of the State Service for Observation (SSO) and Control of Pollution of Natural Environment Objects are divided into three categories by the composition and frequency of observations:

Category I stations (single control stations) are designed for operational control of sea pollution levels. They are usually located in particularly important or permanently polluted sea areas. Observations of pollution and chemical composition of waters are carried out according to a reduced or complete programe. A reduced programe of observations is carried out two to four times a month and a full programe of observations once a month.

Category II stations (single stations or sections) provide systematic information on marine and estuarine water pollution as well as seasonal and inter-annual variability of controlled parameters. The network of these stations covers large areas of sea and estuary water into which waste water enters and from where it may spread. Observations are made in the full programe once a month and once a quarter during the ice age.

Category III stations are designed to obtain systematic information on background pollution levels to study their seasonal and inter-annual variability, as well as to determine the elements of chemical balance. They are located in sea areas with lower pollution levels or in relatively clean waters. Observations are made once a season in a full programe. Background observations are made in areas where contaminants (Cs) can only reach due to their global spread, and in intermediate areas where Cs are received due to regional migration processes.

The category and location of observation stations can be adjusted depending on the dynamics of the level of marine pollution, as well as in connection with the emergence of new objects of control [2].

The reduced program sampled once a decade. Observations usually include determination of the concentration of petroleum hydrocarbons (PH), dissolved oxygen content, pH values and concentrations of one or two priority contaminants specific to the observation area. At the same time, visual observations of sea surface pollution are made.

The full programme is sampled once a month. Observations usually include petroleum hydrocarbons (PH), synthetic surfactants (SSs), phenols, organochlorine pesticides (OCPs), heavy metals (HMs) and area specific contaminants; certain indicators of the marine environment—concentrations of oxygen dissolved in water (O_2), hydrogen sulphide (H_2S), hydrogen ions (pH), alkalinity (Alk), nitrite nitrogen (N–NO_2), nitrate nitrogen (N–NO_3), ammonium nitrogen (N-NH_4), total nitrogen (N_{total}), phosphate phosphorus (P–PO_4), total phosphorus (P_{total}), silicon (Si–SiO_3), as well as elements of the hydrometeorological regime—water salinity (S‰), water and air temperature (T °C), speed and direction of currents and winds, transparency on disk Secchi and color of water, concentration of suspended solids and other parameters [2].

Sampling horizons are determined by depth at the station: up to 10 m—two horizons (surface, bottom); up to 50 m—three horizons (surface, 10 m, bottom); more than 50 m—four horizons (surface, 10 m, 50 m, bottom). If there is a density jump, sampling is also performed at the jump horizon. At deep-water stations, samples are taken at standard hydrological horizons. In expeditionary studies, the set of controlled parameters and the sampling horizons are determined by the work programme.

At the observation points located at the mouth of the estuary in the adjacent river reaches, at a river depth of 1–5 m sampling is carried out on the surface and at the bottom of the river. At a river depth of 5–10 m observations are carried out at the surface, at half the depth and at the bottom, and at a river depth of more than 10 m —at the surface, every 5 m and at the bottom of the river. In points of the third category observations are made 2–4 times a year according to the full programme [2].

Observations of sea and ocean water quality are made on hydrochemical and hydrobiological indicators. The hydrochemical indicators, the definition of which is provided in the framework of a mandatory (complete) observation programme, are presented in Table 12.1. The reduced program of hydrochemical observations

Table 12.1 Parameters to be defined by mandatory (complete) observation programe [2]

Parameter	Units of measure
Petroleum hydrocarbons	mg/dm^3
Dissolved oxygen	mg/dm^3, %
pH	–
Visual observations of surface conditions	–
Chlorinated hydrocarbons, including pesticides	mcg/l
Heavy metals: mercury, lead, cadmium, copper.	mcg/l
Phenols (at surface, at 5, 10, 20 m depth)	mcg/l
SSs (at the surface, 10 m deep, at the bottom)	mcg/l
Additional area-specific parameters	–
Nitrite nitrogen (NO_2^-)	mcg/l
Silicon	mcg/l
Water salinity	Ppm
Water and air temperature	°C
Wind speed and direction	m/sec
Transparency	Score
Excitement (visually)	Score

includes determination of the concentration of oil hydrocarbons, dissolved oxygen, pH and visual observations of the surface of a sea water body.

In the event of new sources of pollution, changes in capacity, composition and forms of discharge, type of water use and other existing conditions, the category of point and list of observed indicators may be changed.

A full programme of observations of seawater quality on hydrobiological indicators includes a study:

- phytoplankton—total biomass, numbers of major groups and species, biomass of major groups and species;
- zooplankton—total biomass, numbers of major groups and species, biomass of major groups and species;
- microbial indicators—total biomass, quantitative distribution of marine microflora indicator groups (saprophyte, oil-oxidizing, xylene oxidizing, phenol oxidizing, lipolytic bacteria), photosynthesis intensity of phytoplankton.

The level of a substance or chemical element in seawater can be determined using a variety of methods and instruments, each characterized by a minimum ingredient detection limit under certain conditions or a concentration level in the analyzed environment (DL = Detection Limit) [2].

In order to describe water quality and compare different water areas by this parameter, it is necessary to use calculated values of water pollution index (WPI), which allow classifying waters of the investigated area to a certain purity class (Table 12.2).

For seawater, four parameters are used to calculate the WPI index, with the obligatory inclusion of dissolved oxygen in this list. Formula for calculating the WPI:

$$\mathbf{WPI} = \sum_{i=1}^{4} \frac{\boldsymbol{C_i}}{\boldsymbol{MACK_i}} \div 4$$

where C_i is the concentration of the three most significant pollutants, the average content of which in the water of the study area exceeded MAC to the greatest extent. The fourth mandatory parameter is the content of oxygen dissolved in water, for which the value in the formula is calculated by dividing the standard (Table 12.3) by the actual content.

Table 12.2 Water quality classes and WPI values [2]

Water quality class		Range WPI of values
Very clean	**I**	WPI $\leq$ 0.25
Pure	**II**	0.25 < WPI $\leq$ 0.75
Moderately polluted	**III**	0.75 < WPI $\leq$ 1.25
Dirty	**IV**	1.25 < WPI $\leq$ 1.75
Dirty	**V**	1.75 < WPI $\leq$ 3.00
Very dirty	**VI**	3.00 < WPI $\leq$ 5.00
Extremely dirty	**VII**	WPI > 5.00

Table 12.3 Norms for dissolved oxygen in water [2]

Diluted oxygen content C (mg/l)	Normal (mg/l)
$6 \leq C$	6
$5 \leq C < 6$	12
$4 \leq C < 5$	20
$3 \leq C < 4$	30
$2 \leq C < 3$	40
$1 \leq C < 2$	50
$C < 1$	60

Since the approved methodology, due to the increased attention to hypoxic conditions, rather describes the ecological acceptability of the water mass for the animals and plants living in it, it seems advisable in the future to abandon the ranking of standards for the content of dissolved O_2 and set one MAC = 6 mgO_2/dm^3 for all cases. This will allow for more accurate assessment of water pollution and the use of WPI for comparative analysis of different water areas.

For cases of extremely high concentrations of individual pollutants in seawater, criteria have been defined for high pollution (HP) and extremely high pollution (EHP) of the marine environment. Boundary conditions for such cases are determined by Order No. 156 of the Head of Roshydromet "On Implementation of the Procedure for Preparation and Presentation of General Purpose Information on Environmental Pollution" dated 31.10.2000 [2].

In addition to contact methods of observations in coastal and open sea areas, remote (space) methods of obtaining information are used. Satellites can provide daily imagery of water areas in the visible, infrared and microwave ranges of electromagnetic radiation. Sounding in the visible range allows obtaining data necessary for the determination of suspended particles, composition and productivity of phyto- and zooplankton, coastal zone state and sea shore dynamics. Infrared and microwave imaging is used to measure ocean surface temperature, detect water salinity, study the thermodynamics of sea ice and other phenomena. Spectral indication is used for qualitative and quantitative analysis of suspensions, determination of chlorophyll in phytoplankton (and indirectly water pollution), detection of oil films on the surface of seas and oceans. As a result, valuable information is accumulated on the areas and intensity of pollution, changes in water properties and transformations of marine ecosystems in time and space.

According to the Order of the Government of the Russian Federation from 10.02.2003 № MK-P9-01617 State Institution "Research Center" Planet "SI" RC "Planet" with the participation of specialists of the Hydrometeorological Center of the Russian Federation, IO RAS and Institute for Space Research (ISR) RAS, since 2003, conducts work on satellite monitoring of pollution (coastal, marine and biogenic) of the water environment of the Russian sector of the Azov-Black Sea basin. At the organization of satellite monitoring world experience of carrying out of similar works, and also specific features of sources of pollution and dynamics of the water environment of the Azov-Black Sea basin are considered.

Over 6500 satellite images of visible, infrared and microwave bands Artificial Earth satellite METEOR-3 M, MONITOR-E, TERRA, AQUA, NOAA, ERS-2, ENVISAT, IRS, QuikSCAT, JASON, TOPEX/Poseidon and METEOSAT-9 were received and processed during the period of observations (http://www.yugmeteo.donpac.ru/monitoring/seasmaps).

In the course of this project, a space monitoring technology was created. Using the developed technology, 12 types of satellite information products are regularly produced, including maps of sea pollution by oil products, water circulation, distribution of phytoplankton and algae, chlorophyll-A concentration, distribution of diffuse attenuation coefficient, sea surface temperature, driving wind, sea level changes, results of automated recognition of water bodies, etc., as well as generalized maps of water environment conditions (os.x-pdf.ru›...kosmicheskiy-monitoring...sredi-azovo...).

In parallel with satellite information, ground measurements are systematized and analysed at meteorological stations in Sochi, Tuapse, Novorossiysk, Kerch, Primorsko-Akhtarsk and Rostov-on-Don, as well as at hydrological stations located at the mouths of the Don, Kuban and Sochi rivers. These data are used for complex analysis of hydrometeorological and environmental conditions.

According to the work data [6] during the monitoring of the natural environment of the Russian sector of the Azov and Black Seas in 2006 more than 1100 space images in the visible, infrared and microwave ranges of the electromagnetic spectrum from 9 satellites specialized in remote sensing of the Earth were received, processed and analyzed. The analysis of the hydrometeorological situation involved the use of ground observation data from meteorological stations in Sochi, Tuapse, Novorossiysk, Anapa, Kerch and Rostov-on-Don, as well as the results of satellite data processing obtained in previous periods of observation. Based on the joint analysis of data received by sensors from different media in different spectral ranges and hydrometeorological information, 13 types of operational satellite information products and generalized maps of the state and pollution of the marine environment [6] were regularly produced.

12.3 Recent Achievements of Satellite Monitoring of the Black Sea in Russia and Prospects for Its Development

In recent decades, the application of space technology has been successfully used as a technical means of observing the processes of anthropogenic impact in the Russian sector of the Black Sea. Under the auspices of the Commission for the Protection of the Black Sea and Cooperation of the Parties to the Bucharest Convention, the Strategic Action Plan (SAP) for minimizing sea pollution by oil was adopted in 2009, with the MONINFO programme based on the application of satellite technologies for detecting sea surface pollution by oil. At the same time, ahead of the EU initiative, the Russian Federation has been implementing the Strategic Action Plan since July 2008 with the participation Information Technology Center "ScanEx" and the Federal State Institution "AMP of the Black Sea", service provider, having access to the Ship Positioning Operating System (SOS), the complex pioneer project "Monitoring of Black Sea Oil Pollution and Environmental Safety of Navigation in the Areas of Intensive Navigation in the Kerch Strait, Water Area of the Novorossiysk Port and on the Approaching Ways to it" [7]. As a result of the introduction of satellite technologies, additional opportunities have opened up for solving the following tasks: (1) Identification of ships involved in unauthorized discharges of oil-containing waters—violators of convention requirements for safety of navigation and environmental protection; (2) technical support in planning and operations in search and rescue areas of the Russian Federation, including with respect to ships that do not supply radio signals; (3) monitoring of intensive navigation zones.

One of the tasks of environmental monitoring carried out by the staff of the Southern Scientific Center of the Russian Academy of Sciences in the Russian sector of the Black Sea is the timely detection and diagnosis of "sea water blooms" caused by the development of potentially toxic algal species, including the most dangerous ones, brought with water ballast ships. With the increasing capabilities and availability of space information systems, the method of measuring chlorophyll concentrations from satellite observations (Marine Portal MHI NASU Remote Sensing Department, access system "http://dvs.net.ua/mp") is being widely used to calculate primary production in different areas of the Black Sea. Due to timely acquisition of space images in March 2008, for the first time in the north-eastern part of the Black Sea it was possible to record the "red tide" caused by the development of Scrippsiella trochoidea (Stein) Balech dinophyte algae species [8, 9]. MODIS optical images taken at that time over the coastal part of the Black Sea helped to track in detail the spatial and temporal distribution of "flowering" waters (web-interface of ScanEx ITCRDC information system).

The obtained results were corrected with chlorophyll cards, demonstrated at the same time by Aqua/MODIS sensors. In 2012, operational observations of the optical properties of the Black Sea surface layer carried out at the MHI of the National Academy of Sciences of Ukraine and regular studies of phytoplankton carried out by the staff of the SSC RAS in the north-eastern part of the Black Sea allowed detecting anomalous in intensity and duration (May–July) "flowering" of water [10]. It was caused by mass development of the nanoplankton species of coccolithophoride Emiliania huxleyi (Lohmann) W. W. Hay and H. P. Mohler. Such a duration and intensity of flowering of coccolithophoride in the Black Sea during the "satellite era" (15 years) has not been observed yet.

Modern satellite technologies make it possible to establish observation of Russian Black Sea water areas exposed to anthropogenic impacts. The tool for this will be a monitoring system, including satellites, measuring equipment on ships and buoys, as well as a center for receiving and processing information. An Aerospace research and development institute for aerospace monitoring is being set up. Scientists are already finalizing an experimental sample of the system, which will soon be tested on the Black Sea coast of the Crimean peninsula and in the Krasnodar Territory. According to scientists, "it is the Black Sea water that is closest to human blood and plasma fluids in terms of composition. Salinity in the upper layer of water here is optimal—1.8%, while in other warm seas the salinity is either too high (3.4–3.6%) or too low, as in the Baltic Sea, for example". There are also many other advantages that are important for tourists, such as the absence of sharks, dangerous jellyfish and other non-hazardous animals.

But the value of these advantages, granted to the Black Sea by nature itself, reduces a great disadvantage coming from man—inordinate human load. This is the discharge of industrial and domestic water, and emergency leakages from ships, hydrocarbon production and a number of other factors. All coastal waters of the Russian part of the Black Sea, including areas near the Crimean peninsula and the city of Sevastopol, suffer from this. The latter, by the way, are especially vulnerable now because of the unfolding construction of transport highways, pipelines and ports, the intensification of offshore development, as well as the construction of cottages near the coast.

In order to be able to monitor the response of the Black Sea coastal ecosystems to human activities, and in the event of a threat to prevent pollution of the sea, a team of scientists from the Research Institute of Aerospace Monitoring "Aerospace" with the participation of specialists from the Marine Hydrophysical Institute of the Russian Academy of Sciences (Sevastopol) and the P. P. Shirshov Institute of Oceanology of the Russian Academy of Sciences is working on a system for monitoring anthropogenic impacts on the shelf zones of the Russian Black Sea coast. This will be a set of special tools aimed at collection, processing and analysis of information important for assessing the state of sea areas. One of the links in the system to be created will be satellites that will provide information on various characteristics of the water environment in coastal areas on-line. In particular, they will provide optical images showing the structure of the sea surface and the surface layer of water, from which one can judge about the color of water, its turbidity, temperature, state of disturbance, presence of oil pollution, film of surfactants. Satellites will also provide radar signals of near-surface wind and detect various contaminants. To receive all these data from space, satellite data reception and primary processing complexes will be used. The use of ground sources for data collection is also envisaged: instruments installed on ships and buoys, as well as equipment deployed on the coast and on hydrophysical platforms. This equipment will make it possible to record water temperature and salinity, wind direction and speed, vertical distribution of temperature, salinity and current velocities, water transparency for the identification of suspended solids and to detect marine pollution.

> "We pay special attention to the integration of the system elements, i.e., the selection of satellites and the equipment installed on them, surface and submersible devices, data reception facilities. An important role is assigned to the development of methods, algorithms and hardware and software tools for processing space, ground and other data, ensuring the identification of anthropogenic impacts on aquatic ecosystems,"—said the project manager, director of the Institute Valery Bondur.—"For this project, the equipment

> of satellites that are already in orbit will be used. If new satellites are launched, our system, as well as all open systems, will perceive the information coming from them for its use in monitoring coastal areas". "In practice, all data collected in this way will come to the information and analytical center. There specialists will already analyze them and identify zones of anthropogenic pollution, as well as monitor the dynamics of these zones. The results will be integrated into an open database from where they will be made available to users via a web interface".
>
> "Today, scientists involved in the project are finalizing the creation of an experimental model of the system, which will combine data from satellite and ground observations, as well as a priori information (all kinds of cartographic data, data on the location of the main sources of pollution in the region, data on the current state of the ecosystem of the water area, etc.). Experiments on development of methods and technologies laid down in the system will start soon. Then will begin testing an experimental sample"—said Academician V. G. Bondur.
>
> "For the experiments selected test sites subject to intense pollution in the area of Sevastopol, near the village of Katsiveli (Southern coast of Crimea), where there is a hydrophysical platform, as well as near the town of Gelendzhik (Krasnodar Territory). If the testing of the system is successful, which, however, scientists have no doubt, recommendations will be developed on how to use the results of monitoring to implement conservation measures—for example, the organization of construction on the coast, or the regime of discharging polluted water into the sea"—said academician V. G. Bondur. In Russia, such an integrated system of regional monitoring of coastal areas to assess human impact on marine ecosystems has not existed so far. Measurements, both remote and contact, are conducted in a fragmented manner and do not allow to see the full picture of the state of a particular water area.

The first preliminary results of setting such a system are presented in the paper [11]. According to the researchers, the project results will be of interest to the Ministry of Natural Resources and Ecology of the Russian Federation, Federal Service for Hydrometeorology and Environmental Monitoring, Ministry of Emergency Situations of Russia and other agencies, as well as shipbuilding, transport, oil and gas producing companies, universities and scientific institutes. The project "Development of methods and creation of an experimental model of the system for monitoring of anthropogenic impacts on the shelf zones of the Black Sea coast of the Russian Federation, including the Crimean peninsula, on the basis of satellite and contact data" is supported by the Federal Target Program "Research and development in priority areas of development of the scientific and engineering complex of Russia for 2014-2020" (http://ecostaff.ru/krym/4477-sistema-monitoringa-chernogo-morya).

Space monitoring data allowed detecting oil products spill in the Black Sea in 146 km from Feodosia in January 2020 (Fig. 12.1).

This was reported to the Federal State Budgetary Institution on January 23, 2020 Scientific Research Center (SRC) for Space Hydrometeorology «Planet». Space monitoring data from Sentinel-1B satellite, received on January 21, 2020, allowed to detect an oil spill in 146 km from Feodosia. SRC Planet identified the object as a film of oil pollution from ships. The area of pollution was 86.1 km^2, the length—55.1 km (Moscow, 24 January—IA Neftegaz.RU). Coastal waters pollution with oil products is one of the main environmental problems in the Black Sea region.

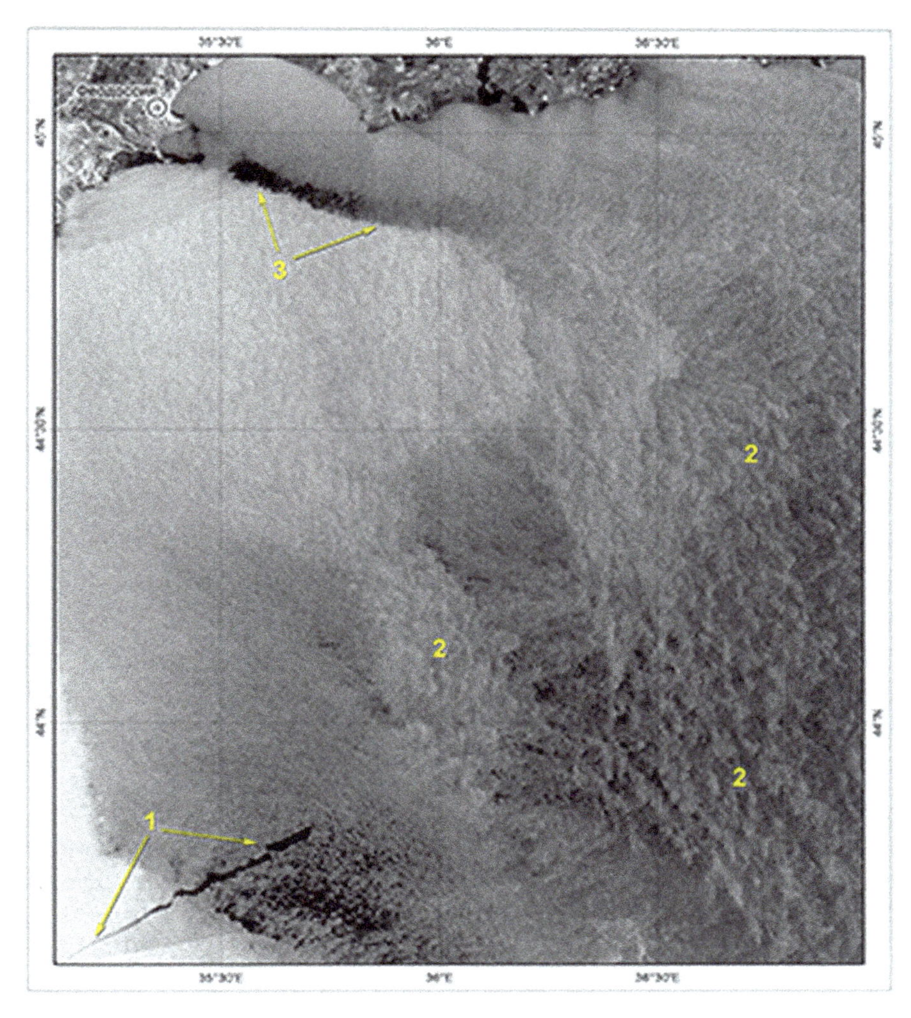

Fig. 12.1 Radar image of the Black Sea off the crimean coast: 1—films of oil spills from ships, 2—effect of atmospheric convection on the excited surface, 3—manifestation of the influence of the atmospheric front on the excited sea surface

12.4 Features, Structure and Objectives of the Black Sea Environmental Monitoring Adopted in the Black Sea Countries (Ukraine)

In the former Soviet Union, there was a system of NOCS (national observatory and control service) that covered all seas, including the Black Sea and the Sea of Azov, which was constantly filled with new tasks (Fig. 12.2). This system included

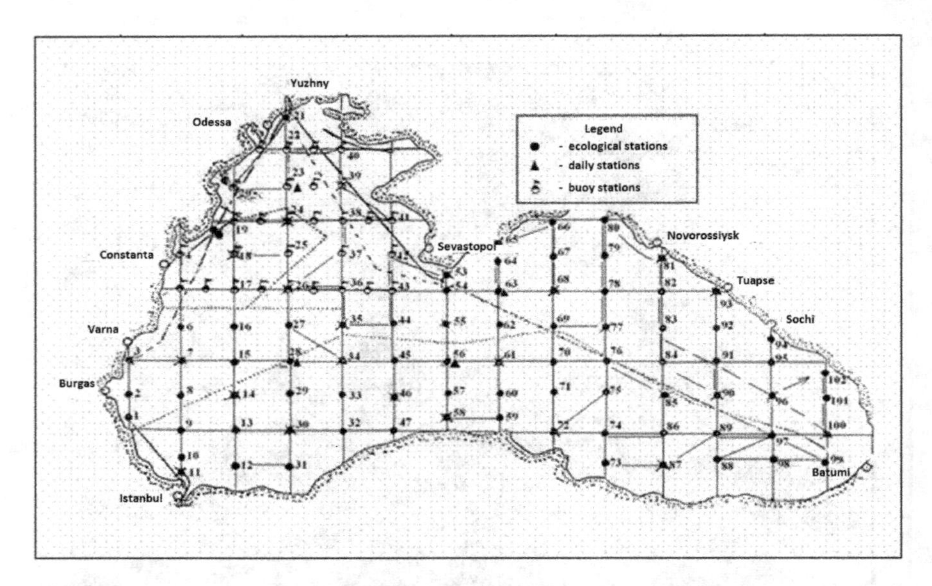

Fig. 12.2 Scheme of ecological surveys of the Black Sea in the USSR (http://www.rusnauka.com/9._EISN_2007/Geographia/21470.doc.htm)

oceanographic, hydrometeorological, hydrobiological, hydrochemical and environmental aspects. This system does not currently exist due to obvious reasons.

In 1992, Ukraine came up with the idea of a three-level monitoring of the water area and intensity of natural use of the sea. This system involves networks of coastal zones, pollution sources and estuaries. This network (several hundred meters) provides for a considerable volume of observations, both standard and special (5–10 days). The regional monitoring area (5–20 km grid) is characterized by a decrease in both standard and special observations. It is carried out in seasonal mode. The open sea zone (grid 50–150 km) of monitoring is characterized by a small number of oceanographic stations with clearly fixed coordinates, a reduction in the set of standard and special observations and a period of 2 times a year. Unfortunately, at present there is no methodology for calculating the optimal density of the observation network, so that neither increase nor decrease of it would be irrational from the economic and scientific point of view.

In recent years, this system has not worked due to the budget depletion in Ukraine. According to the convention, the Odessa Declaration and Strategic Action Plan, the international community has recommended the Black Sea countries move to polygon research. The Strategic Action Plan states that solving the eutrophication problem will require limiting nutrient inputs throughout the Black Sea basin. All countries are invited to agree on a unified approach to water quality assessment and develop a strategy for gradual reduction of pollutant discharges. The riparian countries agree on step-by-step reductions by setting intermediate water quality indicators (https://studbooks.net/887849/ekologiya/strategicheskiy_plan_deystviy_

zaschitu_chernogo_morya). Ukraine, in accordance with the International Black Sea Agreements, has undertaken to carry out environmental monitoring of selected observation points based on rational location, which are chosen on the assessment of the accuracy of restoration of the fields under consideration, scientific optimization, interpolation, taking into account the results of mathematical modeling and model calculations. The research results confirmed the possibility of using research materials for calculation of dynamics and forecasts in short and long-period aspects of the ecosystem.

In (http://www.rusnauka.com/9._EISN_2007/Geographia/21470.doc.htm) an environmental monitoring system that excludes the shortcomings of all previous monitoring concepts is proposed.

This system provides for implementation of three-level monitoring by stations of I, II, and II category, which are characterized by restriction of sampling, analysis of water and bottom sediments for research and objective analysis with the ability to calculate the dynamics, forecast the state of the ecosystem with mathematical modeling and resumption of the international program, work on which was discontinued in 1992 by Ukraine. For Ukraine, 6 sections were fixed in this program: m. B. Fontan–m. Tarkhankut, m. B. Fontan–m. Tarkhankut. Tarkhankut–Snake Island, Chersonesos–Bosporus Ave., Sarych—Inebolu Ave. Kadosh-Unier, Yalta-Batumi.

Subject to Ukraine's compliance with international agreements on the Black Sea, its prestige would significantly increase in the protection and rational use of the Azov-Black Sea basin, which cannot do without environmental monitoring, which makes it possible to determine the current state of the ecosystem and forecast priorities for the conservation and rational use of the Black Sea and the Sea of Azov as an international heritage of global significance.

Marine Environmental Monitoring (MEM) means the environmental monitoring of a marine natural object in this case of the Black and Azov Seas. The objectives of MEM are to assess, diagnose and forecast the state of the marine environment. These tasks are solved on the basis of analysis of data from environmental observations and sources and factors of influence. The objects of marine environmental monitoring are:

- The marine environment within the exclusive maritime economic zone of the Black Sea states in the Black Sea and Azov Sea is the central object of MEM observation.
- Sources of pollution: coastal, marine, river runoff, atmospheric runoff.
- Impactors: hydrometeorological, climatic (seasonal), various types of marine pollution.
- Effects of major pollutants on the physical and chemical parameters of the marine environment, productivity, metabolic processes (ocean-atmosphere, ocean-liquid, ocean-ocean bottom, etc.).

12.5 Proposals to Improve the System of Environmental Monitoring of the Black Sea Under International Projects and Programes

The European and Russian systems of ecological monitoring of the marine environment are fundamentally different.

The Russian system is built on the principle of chemical water analysis (MAC)—maximum allowable concentration of a chemical element in sea water.

The European monitoring system works on the principle "how comfortable a particular organism feels in its habitat. The EU representatives want to implement this monitoring approach on the territory of Russia" (https://www.anapa.info/blogs/39/post_the-project-emblas-another-threat-to-the-black-sea/).

The EMBLAS project "Improving Environmental Monitoring in the Black Sea" (EMBLAS I, II), provides logistical informational and methodological support to the Union of independent states (UIS) countries in fulfilling their obligations under the Bucharest Convention for the Protection of the Black Sea from Pollution in terms of marine environmental monitoring. Following the completion of the first phase of the project and taking into account the results and information obtained, the project is expected to be extended (EMBLAS II), which will include numerous institutional and engineering activities, as well as field cruise studies. Owing to the importance of the objectives, the Black Sea Environmental Monitoring System Performers and Users Meeting was supported by the Ministry of Natural Resources and Environment and Roshydromet.

The aim of the EMBLAS project is to improve the biological and chemical monitoring capabilities of national authorities in Georgia, Russia and Ukraine in the Black Sea. In preparing optimized programmes, it is planned to use the results of the development of practical recommendations in accordance with the requirements of the Water Framework Directive (EU WFD-2000, 2008) and the Marine Strategic Framework Directive (EU MSFD-2008).

EMBLAS I (2013–2014) is a short preparatory phase for the planned extensive technical assistance to the Black Sea region in the preparation and implementation of monitoring programmes. The objective of the second phase of the project, EMBLAS II (2014–2017), is the practical implementation of initiatives developed and planned during the preparatory phase. The project is funded by EU and UNDP.

The objectives of the project are:

1. Improving the quality of monitoring data for the chemical and biological condition of the Black Sea by optimizing the observation systems, taking into account the practical proposals of the EU Directives WFD-2000 and MSFD-2008, as well as the Black Sea Strategic Action Plan (2009);
2. Improving the monitoring capacity of the project partner countries, taking into account the practical recommendations of the EU in the WFD and MSFD Directives and the results of the Black Sea Diagnostic Reports I and II study.

In the first and second phases of the EMBLAS project, the following actions are envisaged:

- Analysis of national monitoring systems and access capabilities to monitoring data;
- Support to countries in fulfilling their obligations under the Bucharest and other international conventions;
- Development and promotion of programs for cost-effective and harmonized biological and chemical monitoring of the marine environment in accordance with international agreements, as well as WFD and MSFD;
- Assessment of countries' needs for laboratory infrastructure, equipment, training and teaching materials;
- Development and implementation of a training program on monitoring methods and quality assurance of results in accordance with ISO 17025;
- Preparation of the Joint Black Sea Surveys methodology and its implementation to prepare an assessment of the state of the Black Sea off the coast;
- Development and creation of the Black Sea Water Quality Database on a web basis.

On March 12–13, 2015 in Sochi, within the framework of the first phase of the project "Improvement of the environmental monitoring of the Black Sea" (EMBLAS, www.emblasproject.org) a meeting of organizers, performers and consumers of information of the environmental monitoring system of the Black Sea was held. At this meeting they heard reports on the current status of marine environmental monitoring in Russia, Georgia and Ukraine, regulatory international and national documents in this area, proposals for optimization and improvement of monitoring programmes, taking into account modern achievements and partial changes in the goals and objectives of monitoring systems (http://azovinform.ru/news/6/).

The meeting was attended by 45 people, including representatives of the Ministry of Natural Resources and Ecology, Roshydromet, Rosprirodnadzor, Rosrybolovstvo, the Institute of Oceanology of the Russian Academy of Sciences named after the P. P. Shirshov. The meeting was attended by 45 people, including representatives of the Ministry of Natural Resources and Ecology, RosHydroMet, Rosprirodnadzor, Rosrybolovstvo, the Institute of Oceanology of the Russian Academy of Sciences named after Shirshov, "Scientific and Production Association" Typhoon of All-Russian Research Institute of Information, State Oceanographic Institute "SOIN", Sevastopol branch of SOIN, Institute of Biology of the South Seas (InBSSs), VNIRO (Moscow), YugNIRO (Kerch), MHI (Sevastopol), representatives of the Regional Centres and Departments for Hydrometeorology and Environmental Monitoring of Sochi, Krasnodar, Tuapse, Rostov, as well as representatives of other organizations. At the invitation of the EMBLAS Project Coordinator in the Russian Federation, A. N. Korschenko, the meeting was attended by the staff of the Crimean Department of HydroMeteoService Parfenova Veronika—Head of the Laboratory for

Environmental Pollution Monitoring in Yalta and Furnik Tatiana—oceanologist of the 1st category of marine hydrometeorological station of Yalta.

A wide discussion was held on proposals to change the monitoring in order to develop the optimal composition and procedure of observations, practical steps to collect, store and process monitoring data, as well as information dissemination and international exchange. During the meeting, the coordinator of the EMBLAS Project in the Russian Federation, A. N. Korschenko (SOIN, Moscow), proposed a program "pilot monitoring in the coastal part of the Russian Black Sea coast. Category I station Public state service of observation and control (SSOC) No 139806103, located in the water area of Yalta seaport, where the Yalta Chemical Laboratory conducts annual sampling for sea water quality, can be included in this Project" (http://meteo.crimea.ru/?p=2248). Also at this meeting it was announced that the main contaminant of the Black Sea waters is the European rivers flowing into the Black Sea, and the main source is the Danube River.

All financial costs of the Meeting were covered by the Secretariat of the UNDP and EU international project "Improvement of Environmental Monitoring in the Black Sea" (EMBLAS-1), which aims to improve the ability of national authorities in Georgia, Russia and Ukraine to implement internationally optimized environmental monitoring programmes for the Black Sea.

As a result of the meeting, a special questionnaire was developed and sent to all major Black Sea monitoring participants in Georgia, Russia and Ukraine to collect information on the current status of monitoring systems, storage and analysis of collected data. Using the data from the questionnaire and other available sources of information, an overview of the monitoring programes of the three countries (Black Sea Diagnostic Report II) was prepared, totalling approximately 400 pages. The review contains not only the analysis of the current situation, but also recommendations for improving the methodology of monitoring, management and analysis of the collected data, infrastructure and equipment for its implementation (http://www.oceanography.ru/index.php/2013-05-24-16-12-00/2013-11-21-09-35-22/396-emblas).

The review also addresses issues of assessment of water quality and ecological status of marine ecosystems (WQ/GES classification). An additional document, Compliance Indicators, has been drafted to assess countries' compliance with their international obligations in relation to the Black Sea. To improve chemical monitoring, a training programme on chemical analysis techniques was prepared, and the first training workshop for participants from Georgia, Russia and Ukraine was held in June 2014 in Batumi.

The project experts prepared four methodological guidelines in English on micro-, meso-, macro-zooplankton and macrophytobenthos monitoring, aimed at evolving a common methodological framework for monitoring these biota groups, improving the accuracy of its results and data comparability. Within the framework of the project, a methodology for joint Black Sea monitoring voyages (Joint Black Sea Surveys) has been developed. To date, a prototype of the Black Sea Water Quality Database and the Black Sea Information System has also been developed.

The database will be supported by the Black Sea Commission secretariat to improve the exchange of information between national and international bodies.

As mentioned above, one of the most important tasks of the project is to improve the national Black Sea monitoring programs in Georgia, Russia and Ukraine and to develop them in accordance with international requirements and obligations of the countries. This unit of project work is still under development. An important stage of its implementation will be workshops in each of the three countries, aimed at discussing ways to develop modern national Black Sea monitoring systems. It is planned that all main monitoring participants in each country, as well as employees of central and regional authorities, representatives of the general environmental community and other consumers of marine environmental monitoring system information are invited to participate in these seminars (https://www.anapa.info/blogs/39/post_the-project-emblas-another-threat-to-the-black-sea/).

Within the framework of the international EMBLAS project "Improving Environmental Monitoring in the Black Sea" (No. 88460: Improving Environmental Monitoring in the Black Sea, Phase 2—EMBLAS-II) funded by the United Nations Development Programe (UNDP) and the European Community (EC), an expedition to the coastal waters of the Adler-Sochi region was conducted on 15 November. A total of 8 stations with depths from 6.7 to 370 m were carried out, 22 samples were taken from the surface and intermediate layers to a depth of 58 m. The aim of the expedition was to obtain data on the ecological state of the marine environment in the area between the mouths of the Mzymta and Sochi rivers. The results of meteorological and hydrological observations were made and recorded, standard hydrochemical parameters and concentrations of various pollutants in water were determined. The content of heavy metals (HM) and persistent organic pollutants (POPs) in the tissues of fish and bivalve molluscs (mussels) selected in the Sochi port water area were also determined. The species composition, abundance and biomass of phytoplankton and concentration of photosynthetic pigments were determined.

Assessment of water district Sochi-Adler. According to the results of the expedition, in 2017 the level of pollution of coastal waters of the Greater Sochi area between the estuaries of the Mzymta and Sochi rivers increased compared to the previous year, but remained below 2015. According to the calculated complex index of water pollution of the WPI in 2017 (0.73), water is classified as "clean" (see Table 2). The average concentration of most of the normalized pollutants was below the standards set for marine waters. At the same time, the maximum concentration in individual samples exceeded MAC for oil hydrocarbons (up to 1.14 MAC), iron (3.5 MAC), lead (3.3 MAC), suspended solids (2.3 MAC), synthetic surfactants (SSs) (5.4 MAC). The greatest maintenance of the easily oxidized organic substance defined on БПК5, made 1.1 MAC. Dissolved mercury was not detected in the waters of the region. In 2017, the waters of the port of Sochi ("moderately polluted") were the most polluted in comparison with estuarine sections of the rivers Sochi, Khosta and Mzymta and open sea waters ("clean") (see

Table 2). There were no significant differences in the composition of contaminants —lead, iron and organic substances as per BOD5.

The water pollution index of the entire water area from Mzymta to Sochi was high: 58%, since 7 parameters out of 12 were higher than MAC (BOD_5, PO_4, SS, Fe, Pb, petroleum products PP and suspended matter SM). Waters of the region are characterized by a single frequency of excess of MAC (less than 10%) for oil hydrocarbons (1.6%, one sample out of 64), suspended solids (5.7%), Five samples from 88), BOD_5 (3.1%, two samples from 64), SS (3.1%, two samples from 64), mineral phosphorus (4.7%, three samples from 64) and iron (28.1%, 18 samples from 64) and lead (28.1%, 18 samples from 64). The MAC exceedance factor was average (2–10 times) for PO_4, SM, SS, Fe and Pb and low (1–2 times) for BOD_5 and PP. No significant changes in seawater quality have been observed in the last few years. The overall level of pollution is insignificant, and waters have been described as "clean" and "moderately polluted" (see Table 12.2). The condition of the area's waters is assessed as stable over the years.

As part of the EMBLAS-II project, another expedition was conducted in August 2017. According to Yaroslav Slobodnik, EMBLAS-II project manager from Ukraine: "Within the framework of the EMBLAS-II expedition the largest research ship of the European Union, Mare Nigrum, travelled in the waters of Ukraine and Georgia. On the way we collected samples of water, with which we conducted many experiments. We checked each sample for 2100 dangerous compounds. Fortunately, not everything was found. But the concentration of some of them is killing the flora and fauna of the Black Sea".

According to Viktor Komorin, Director of the Ukrainian Research Center for Marine Ecology (UkrCEM), "ecologists have high hopes for this expedition, which is the second one supported by the EU within the framework of the EMBLAS-II project". According to him, "the task of the expedition is to assess the state of deep waters of the Black Sea using European methods". One of the tasks of the expedition was emphasized by Viktor Comorin "to find out how much man and natural factors influence the state of the sea". "An equally important task is to study sea pollution not only on the sea surface, but also in bottom sediments". Thanks to the EMBLAS-II project, it is possible to study the entire Black Sea from Odessa to Batumi in order to understand the state of the sea (https://www.unian.net/ecology/naturalresources/2104354-v-odesse-nachalas-mejdunarodnaya-nauchnaya-ekspeditsiya-po-issledovaniyu-ekologii-c).

Thus, the basis for effective monitoring of the water area of the Azov-Black Sea basin should be formed by studies of water pollution levels, obtaining information on changes in the contours of the coastline and estuaries of the rivers flowing into the sea as a result of economic activity, an inventory of sources of anthropogenic pollution of coastal waters, toxicological studies of coastal waters, soil and bioresources.

12.6 Conclusion

The analysis of features, structure and objectives of environmental monitoring of the Black Sea, its means and methods adopted in the Russian Federation is presented. The standards adopted by RoskomHydromet on the methods of analysis, schemes of marine sampling stations location, analyzed parameters, peculiarities of the State observation network formation are described.

Features of satellite monitoring of the Azov-Black Sea basin are considered. Information on the technology of space monitoring of natural environment in the Russian sector of the Azov and Black Seas is presented, and possibilities of joint analysis of data, received by sensors from different carriers in different spectral range, and hydrometeorological information as a source of operational satellite information products in the form of generalized maps of marine environment condition and pollution are shown.

The latest achievements of satellite monitoring of the Black Sea in the Russian Federation and its development prospects are analyzed. Additional possibilities of satellite technologies implementation in solving a number of environmental tasks are listed. Expanded possibilities and availability of space information systems for calculation of primary production in different areas of the Black Sea are shown on specific examples.

A new system for monitoring of anthropogenic impacts on the shelf zones of the Russian Black Sea coast is under consideration, which is being set up by scientists from the Aerospace Research Institute of Aerospace Monitoring with the participation of specialists from the Marine Hydrophysical Institute of the Russian Academy of Sciences (Sevastopol) and P. P. Shirshov Institute of Oceanology of the Russian Academy of Sciences (Moscow).

The changes that have taken place in the structure and tasks of environmental monitoring of the Black Sea adopted in the Black Sea countries on the example of Ukraine after 1992 are analyzed. The system of three-level monitoring, proposed in 1992 by Ukraine, according to which the regional monitoring zone (grid 5–20 km) is characterized by a reduction of standard and special observations, is carried out in seasonal mode. The open sea zone (50–150 km grid) is characterized by the presence of a small number of oceanographic stations. However, in recent years this system has not been working due to material difficulties in Ukraine.

The differences in the European and Russian systems for environmental monitoring of the marine environment are considered. Two stages of implementation of the International EMBLAS Project "Improvement of Environmental Monitoring in the Black Sea" are analyzed, the main objective of which is to improve the capacity of national authorities in Georgia, Russia and Ukraine to carry out biological and chemical monitoring in the Black Sea. Providing extensive technical assistance to the Black Sea region in preparation and implementation of monitoring programmes and improvement of the Black Sea environmental monitoring system.

The results of the meeting of the organizers, implementers and consumers of the Black Sea environmental monitoring system information held in Sochi in 2015 as part of the first phase of the project "Improvement of environmental monitoring of the Black Sea" (EMBLAS, www.emblasproject.org) are described. Proposals to change the monitoring were widely discussed in order to develop the optimal composition and procedure of observations, practical steps to obtain, store and process monitoring data, as well as information dissemination and international exchange.

Within the framework of the international EMBLAS project "Improvement of methods for environmental monitoring of the Black Sea", funded by the United Nations Development Programe (UNDP) and the European Community (EC), expedition research was carried out in the Black Sea in 2017. In November, an expedition was carried out in the coastal waters of the Adler-Sochi region. In the framework of the EMBLAS-II project, another expedition was conducted in August 2017. The objective of the expedition was to assess the state of deep waters of the Black Sea and coastal areas of Ukraine and Georgia using European methods.

References

1. Panov BN (2018) Ecological condition of the Azov-Black Sea region. Methodical instructions on independent work and on performance of control work for students of a direction of preparation 05.04.06 "Ecology and nature management", Kerch Seal Technological University, Kerch, 19 p
2. Korshenko AN (ed) (2018) The quality of sea waters by hydrochemical indicators. Yearbook "seawater quality on hydrochemical indicators", Science, Moscow, 220 p
3. Order No 156 of the Head of RosHydroMet "on enactment of the Procedure for preparation and submission of general-purpose information on environmental pollution" of 31.10.2000
4. http://www.yugmeteo.donpac.ru/monitoring/seasmaps
5. os.x-pdf.ru›…kosmicheskiy-monitoring…sredi-azovo
6. Krovotintsev VA, Lavrova OY, Mityagina MI, Ostrovsky AG (2009) Space monitoring of natural environment in the Azov-Black Sea Basin. IKI RAS conference, Moscow, (DVDROM), pp 295–303
7. Matishov GG, Matishov DG, Berdnikov SV, Kovaleva GV, Vikrischuk AV (2011) Risks of the geological exploration and oil production projects realization in the Black Sea sulfide zone conditions (in Russian) 7(1):59–64
8. Yasakova ON, Berdnikov VS (2008) Unusual flowering as a result of the dinophyte algae development in the water area of the Novorossiysk bay of the Black Sea in March 2008 (in Russian). Ecol J 7(4):98
9. Yasakova ON, Berdnikov VS (2009) Monitoring of "red tides" in the Black Sea. Earth Space Most Effective Solutions 3:1214
10. Yasakova ON, Stanichnyi SV (2012) Abnormal flowering *Emiliania huxleyi* (Prymnesiophyceae) in 2012 in the Black Sea. Marine Ecol J XI(4):64
11. Bondur VG, Ivanov VG, Vorobiev VE, Dolotov VA, Zamshin VV, Kondratyev SI, Li ME, Malinovsky VV (2020) Terrestrial and space monitoring of the anthropogenic influences on the coastal zone of the crimean peninsula (in Russian). Marine Hydrophys J 36(1):103–115. https://doi.org/10.22449/0233-7584-2020-1-103-115
12. http://ecostaff.ru/krym/4477-sistema-monitoringa-chernogo-morya

13. https://studbooks.net/887849/ekologiya/strategicheskiy_plan_deystviy_zaschitu_chernogo_morya
14. https://www.anapa.info/blogs/39/post_the-project-emblas-another-threat-to-the-black-sea/
15. http://meteo.crimea.ru/?p=2248
16. http://www.oceanography.ru/index.php/2013-05-24-16-12-00/2013-11-21-09-35-22/396-emblas
17. https://www.unian.net/ecology/naturalresources/2104354-v-odesse-nachalas-mejdunarodnaya-nauchnaya-ekspeditsiya-po-issledovaniyu-ekologii-c

CPSIA information can be obtained
at www.ICGtesting.com
Printed in the USA
BVHW012354081220
595246BV00001B/11

9 783030 61894